Mastering Telecommunications Engineering: A Comprehensive Guide

Henry

Copyright © [2023]

Author: Henry

Title: Mastering Telecommunications Engineering: A Comprehensive Guide

This book is a self-published work by the author Henry

ISBN:

TABLE OF CONTENTS

Chapter 3: Telecommunication Networks

Chapter 4: Telecommunication Systems   58

Chapter 5: Telecommunication Technologies

Chapter 1: Introduction to Telecommunications Engineering

The Evolution of Telecommunications

Telecommunications, a crucial aspect of our modern lives, has undergone a remarkable evolution over the years. From the early days of telegraphs and landline telephones to the wireless networks and high-speed internet connections of today, the field of telecommunications has advanced at an astonishing pace. This subchapter explores the fascinating journey of telecommunications, its milestones, and the impact it has had on society.

The origins of telecommunications can be traced back to the invention of the telegraph in the early 19th century. Developed by Samuel Morse, the telegraph allowed the transmission of coded messages over long distances using electrical signals. This breakthrough technology revolutionized long-distance communication and laid the foundation for future advancements.

The next major leap in telecommunications came with the invention of the telephone by Alexander Graham Bell in 1876. This revolutionary device enabled voice communication, replacing the need for written messages. The telephone network quickly expanded, connecting people across cities and countries, and became an essential tool for businesses and individuals alike.

In the 20th century, the advent of radio broadcasting and television further transformed the telecommunications landscape. Radio allowed for the transmission of audio signals wirelessly, while television added the visual element, bringing live images and entertainment into people's homes. These technologies not only revolutionized entertainment but also played a crucial role in disseminating news and information on a global scale.

The digital revolution of the late 20th century marked another significant turning point in telecommunications. The development of digital technologies allowed for the encoding and transmission of data in binary form, leading to the birth of the internet and mobile communications. The internet revolutionized the way we communicate, access information, and conduct business. It opened up a world of possibilities, connecting people from different corners of the globe and enabling instant communication and collaboration.

Today, we are witnessing the era of wireless communication, with the widespread adoption of mobile phones and the emergence of 5G networks. These advancements provide faster and more reliable connections, paving the way for innovations such as the Internet of Things (IoT) and smart cities. The evolution of telecommunications has not only facilitated personal communication but has also transformed industries, healthcare, transportation, and many other sectors.

In conclusion, the evolution of telecommunications has been a remarkable journey filled with groundbreaking inventions and technological advancements. From the telegraph to the internet, this field has shaped the way we communicate, connect, and interact with the world. As electrical engineers, understanding the evolution and current state of telecommunications is crucial for designing and developing the technologies of the future.

Importance of Telecommunications Engineering

The Importance of Telecommunications Engineering

In today's interconnected world, telecommunications engineering plays a crucial role in enabling communication and information exchange across vast distances. From the simple act of making a phone call to the complex network infrastructure supporting the internet, telecommunications engineering is at the heart of it all. This subchapter will delve into the significance of telecommunications engineering, focusing on its relevance to the field of electrical engineering.

Telecommunications engineering is a multidisciplinary field that combines elements of electrical engineering, computer science, and communication technology. Its primary objective is to design, develop, and maintain systems that facilitate the transmission of information, whether it be through wired or wireless means. These systems include telephone networks, satellite communication systems, fiber-optic cables, and data networks.

For electrical engineers, understanding the principles and applications of telecommunications engineering is of utmost importance. The field not only expands their knowledge base but also opens up a plethora of career opportunities. With the rapid advancement in technology, the demand for skilled telecommunications engineers is on the rise.

One of the key reasons why telecommunications engineering is vital in the realm of electrical engineering is its role in the development of cutting-edge communication systems. As electrical engineers, it is crucial to stay updated with the latest trends and innovations in telecommunications technology. This includes advancements in wireless communication, data transmission protocols, network security, and signal processing techniques.

Furthermore, telecommunications engineering is essential for the seamless integration of various electrical systems. From power distribution networks to smart grid systems, telecommunications engineering ensures efficient communication between different components, leading to optimized performance and increased reliability.

Moreover, the importance of telecommunications engineering extends beyond the realm of electrical engineering. It has a significant impact on various industries, including healthcare, finance, transportation, and entertainment. Telecommunications engineers are instrumental in developing communication solutions that enhance patient care, enable secure financial transactions, improve transportation systems, and deliver high-quality entertainment experiences.

In conclusion, telecommunications engineering is a critical field that holds immense importance in the realm of electrical engineering. It enables efficient communication and information exchange, facilitates technological advancements, and opens up new career opportunities. By mastering the principles of telecommunications engineering, electrical engineers can make significant contributions to a wide range of industries and shape the future of communication technology.

Overview of Telecommunications Systems

Telecommunications systems play a crucial role in today's interconnected world, enabling the seamless exchange of information across vast distances. From voice calls to video conferencing, these systems have revolutionized the way we communicate and conduct business. This subchapter provides a comprehensive overview of telecommunications systems, exploring their fundamental concepts and technologies.

Telecommunications systems are complex networks of interconnected devices that facilitate the transmission of information. These systems encompass a wide range of technologies, including wired and wireless communication, satellite systems, and fiber optics. Understanding the basics of these systems is essential for professionals in the field of electrical engineering.

The subchapter begins by introducing the fundamental principles of telecommunications systems. It explains how data and information are transmitted through various channels, such as copper wires, optical fibers, and radio waves. The subchapter also covers the concept of modulation, which is the process of encoding information onto a carrier signal for transmission.

Next, the subchapter delves into the different components of a telecommunications system. It explains the role of transmitters, receivers, and amplifiers in signal transmission and reception. It also covers the importance of protocols and standards in ensuring compatibility and interoperability between different devices and networks.

The subchapter then explores the various types of telecommunications systems, including landline networks, mobile networks, and satellite systems. It examines the advantages and limitations of each system, highlighting their applications in different contexts. Additionally, it discusses emerging technologies such as 5G, Internet of Things (IoT), and virtual reality, which are shaping the future of telecommunications.

Furthermore, the subchapter addresses the importance of security in telecommunications systems. It discusses encryption and authentication techniques used to protect sensitive information from unauthorized access and data breaches. It also touches upon the challenges and ethical considerations associated with privacy in today's interconnected world.

In conclusion, this subchapter provides a comprehensive overview of telecommunications systems, covering their fundamental concepts, technologies, and applications. It is a valuable resource for professionals in the field of electrical engineering, as well as anyone interested in understanding the intricate workings of modern communication networks. Whether you are a student, a researcher, or a technology enthusiast, this subchapter will equip you with the knowledge needed to navigate the dynamic and ever-evolving world of telecommunications.

Chapter 2: Fundamentals of Telecommunications Engineering

Basic Concepts in Telecommunications

Telecommunications is a vast field that encompasses the transmission of information, such as sound, images, and data, over long distances using various technologies. In this subchapter, we will explore the fundamental concepts that form the foundation of telecommunications engineering. This chapter is part of the book "Mastering Telecommunications Engineering: A Comprehensive Guide," which aims to provide a comprehensive understanding of the subject matter to a diverse audience, including electrical engineering enthusiasts and professionals.

The subchapter begins by introducing the concept of telecommunications and its significance in today's interconnected world. It explains how telecommunications enables the exchange of information across long distances, enabling seamless communication between individuals and devices. The subchapter emphasizes that telecommunications is not limited to traditional telephone networks but also includes various modern technologies such as wireless communication, satellite systems, and the internet.

Next, the subchapter delves into the basic components and principles of telecommunications systems. It provides an overview of the key elements, such as transmitters, receivers, communication channels, and protocols, that work together to ensure effective communication. The subchapter also covers crucial concepts like modulation, demodulation, encoding, decoding, and multiplexing, which are essential for signal transmission and reception.

Furthermore, the subchapter addresses the various transmission media used in telecommunications. It explores wired media, such as copper wires and fiber optics, as well as wireless media, including radio waves and microwaves. It highlights the advantages and limitations of each medium and explains how they are used in different applications.

Additionally, the subchapter introduces the concept of network architecture and its importance in telecommunications. It discusses the different types of networks, such as local area networks (LANs), wide area networks (WANs), and the internet. The subchapter explains how these networks are interconnected and governed by protocols to enable seamless communication between devices worldwide.

Lastly, the subchapter touches upon emerging technologies and trends in telecommunications. It provides a glimpse into cutting-edge concepts like 5G, Internet of Things (IoT), and virtual reality (VR), which are revolutionizing the way we communicate and interact with our surroundings.

In conclusion, this subchapter on basic concepts in telecommunications serves as a comprehensive introduction to the field. It equips the readers, including electrical engineering enthusiasts and professionals, with a solid understanding of the fundamental concepts, principles, and technologies that underpin telecommunications engineering. Whether you are a beginner or an experienced professional, this subchapter will provide valuable insights into the world of telecommunications and its limitless possibilities.

Signals and Waves

In the vast field of electrical engineering, understanding signals and waves is crucial. Whether you are a seasoned professional or a student just starting out, this subchapter will provide you with a comprehensive understanding of the fundamental concepts of signals and waves.

Signals and waves are the backbone of modern telecommunications systems, serving as the carriers of information. They can be found in various forms, ranging from simple audio signals to complex radio waves. By delving into the world of signals and waves, you will develop a solid foundation to comprehend and analyze the functioning of advanced telecommunication systems.

This subchapter will commence by introducing the concept of signals. You will explore the different types of signals, including analog and digital signals, and understand their characteristics and applications. Additionally, you will learn about signal processing techniques, such as modulation, demodulation, filtering, and encoding. These techniques play a vital role in transmitting and receiving signals efficiently.

Moving forward, the subchapter will delve into the fascinating world of waves. You will gain a deep understanding of the properties of waves, including frequency, amplitude, and phase. Additionally, you will explore the different types of waves, such as electromagnetic waves, sound waves, and radio waves. Understanding the behavior and propagation of waves is essential for designing and implementing various telecommunication systems.

To enhance your comprehension, practical examples and case studies will be provided throughout the subchapter. These real-life scenarios will demonstrate the application of signals and waves in various telecommunication systems, including cellular networks, satellite communications, and fiber optics.

By the end of this subchapter, you will have a solid grasp of signals and waves, enabling you to analyze, design, and troubleshoot telecommunication systems with confidence. Whether you are an electrical engineering student seeking to excel in your studies or a professional looking to enhance your skills, this subchapter will serve as an invaluable resource.

Mastering Telecommunications Engineering: A Comprehensive Guide aims to equip every reader, regardless of their level of expertise, with the knowledge and skills needed to excel in the field of electrical engineering. With a strong foundation in signals and waves, you will be well-prepared to tackle the challenges of the ever-evolving telecommunications industry.

Modulation Techniques

In the vast world of telecommunications engineering, the efficient transmission of information is crucial. This is where modulation techniques come into play. Modulation is the process of modifying a carrier signal to carry information, enabling the successful transfer of data over long distances. In this subchapter, we will explore various modulation techniques that are fundamental to the field of electrical engineering.

One of the most widely used modulation techniques is amplitude modulation (AM). AM involves varying the amplitude of a high-frequency carrier signal in proportion to the instantaneous amplitude of a low-frequency information signal. This technique allows the transmission of audio signals over radio waves, making it indispensable for broadcasting and two-way communication systems.

Frequency modulation (FM) is another popular modulation technique. Unlike AM, FM involves varying the frequency of the carrier signal based on the instantaneous amplitude of the input signal. FM is commonly used in broadcast radio and television systems due to its superior resistance to noise interference. It provides a higher signal-to-noise ratio, resulting in improved audio and video quality.

Phase modulation (PM) is a technique that alters the phase of the carrier signal according to the input signal. PM is often used in digital communication systems to transmit data. By manipulating the phase of the carrier signal, PM enables the encoding and decoding of digital information, making it an essential component of modern telecommunications.

Quadrature amplitude modulation (QAM) is a modulation technique that combines both amplitude and phase modulation. By varying both the amplitude and phase simultaneously, QAM can transmit multiple bits of information in a single symbol, increasing the data transmission rate. QAM is widely used in digital communication systems such as cable modems and wireless networks.

In summary, modulation techniques play a vital role in electrical engineering, enabling the successful transmission of information over long distances. From the fundamental AM and FM techniques to the more advanced PM and QAM methods, mastering these techniques is essential for anyone involved in telecommunications engineering. Whether you are a student, professional, or simply interested in the field, understanding modulation techniques is the key to unlocking the potential of modern communication systems.

Multiplexing Methods

In the realm of telecommunications engineering, the ability to transmit multiple signals simultaneously over a single channel is of utmost importance. This process, known as multiplexing, has revolutionized the field by increasing the efficiency and capacity of communication systems. In this subchapter, we will delve into the various multiplexing methods employed in electrical engineering, shedding light on their underlying principles and applications.

Time Division Multiplexing (TDM) is one of the fundamental techniques used to achieve multiplexing. It divides the available time slots into smaller intervals, assigning each slot to a specific signal. By rapidly switching between these signals, TDM enables multiple signals to share a common transmission medium. This method is widely utilized in various applications, ranging from telephone networks to digital television broadcasting.

Frequency Division Multiplexing (FDM) is another widely used technique. It involves allocating different frequency ranges to separate signals, allowing them to coexist on a shared channel. FDM is commonly employed in radio and television broadcasting, where different channels are assigned specific frequency ranges. By utilizing the full bandwidth of the channel, FDM maximizes the number of signals that can be transmitted simultaneously.

Statistical Time Division Multiplexing (STDM) is a more advanced form of multiplexing that dynamically allocates time slots based on the actual data traffic. This method intelligently assigns time slots to signals with higher data rates, ensuring efficient utilization of the available bandwidth. STDM is commonly used in modern data networks, where the demand for bandwidth varies over time.

Code Division Multiplexing (CDM) is a multiplexing technique that assigns a unique code to each signal, allowing them to be transmitted simultaneously. By utilizing different code sequences, CDM enables multiple signals to overlap in both time and frequency domains. This method is widely employed in wireless communication systems, such as 3G and 4G networks.

Orthogonal Frequency Division Multiplexing (OFDM) is a multiplexing method that divides the channel into multiple subcarriers, each carrying a different signal. By orthogonalizing these subcarriers, OFDM eliminates interference between them, enabling high-speed data transmission. OFDM is commonly used in modern digital communication systems, including Wi-Fi and 5G networks.

Understanding the principles and applications of these multiplexing methods is crucial for electrical engineering professionals working in the field of telecommunications. By efficiently utilizing the available resources, multiplexing methods enable us to transmit an ever-increasing amount of data over limited channels. This subchapter provides a comprehensive overview of the various multiplexing techniques, empowering engineers to design and optimize communication systems that meet the demands of the modern world.

Whether you are a student, researcher, or industry professional, this subchapter is a valuable resource for gaining a deep understanding of multiplexing methods in telecommunications engineering. By mastering these techniques, you will be well-equipped to contribute to the advancements and innovations in the field of electrical engineering, and play a crucial role in shaping the future of telecommunications.

Transmission Media

Transmission media refers to the physical pathways that enable the transfer of data and information from one point to another in a telecommunications network. These pathways play a crucial role in ensuring reliable and efficient communication between devices and systems. In this subchapter, we will explore the various types of transmission media commonly used in electrical engineering and their applications in telecommunications.

One of the most widely used transmission media is copper wire. Copper wires have been traditionally used for transmitting electrical signals due to their low cost, flexibility, and ease of installation. They are commonly used in telephony networks and local area networks (LANs). However, copper wires have limitations in terms of bandwidth and distance, which led to the development of alternative transmission media.

Fiber optic cables have revolutionized the field of telecommunications. These cables use strands of glass or plastic fibers to transmit data through the use of light pulses. Fiber optic cables offer several advantages over copper wires, including high bandwidth, low signal loss, and long-distance transmission capabilities. They are extensively used in long-haul telecommunications networks, internet backbones, and high-speed data transmission applications.

Wireless transmission media have gained significant popularity in recent years. They allow communication without the need for physical cables, providing flexibility and mobility in various applications. Wireless transmission media include radio waves, microwaves, and infrared signals. They are widely used in wireless communication systems, such as cellular networks, satellite communication, and Wi-Fi networks.

In addition to these commonly used transmission media, there are specialized transmission media for specific applications. For instance, coaxial cables are used for high-frequency applications, such as cable television and broadband internet. Twisted pair cables, on the other hand, are commonly used in Ethernet networks for short-distance data transmission.

Choosing the appropriate transmission media for a particular application requires careful consideration of factors such as bandwidth requirements, distance limitations, cost, and environmental factors. Electrical engineers play a crucial role in designing and implementing efficient and reliable telecommunication networks by selecting the most suitable transmission media.

In conclusion, transmission media form the physical pathways that enable the transfer of data and information in telecommunications networks. Copper wires, fiber optic cables, and wireless transmission media are among the commonly used types of transmission media in electrical engineering. Each type has its own unique characteristics and applications. Understanding the capabilities and limitations of different transmission media is essential for designing and implementing effective telecommunication systems.

Telecommunications Standards and Protocols

In the rapidly evolving field of telecommunications, standards and protocols play a crucial role in ensuring seamless communication between various devices and networks. This subchapter delves into the intricacies of telecommunications standards and protocols, providing a comprehensive understanding of their significance in the realm of electrical engineering.

Standards serve as a common language that enables different manufacturers and service providers to develop compatible equipment and services. They provide a framework for interoperability, ensuring that devices from different vendors can communicate with each other effectively. Without standards, the telecommunications industry would be chaotic, with incompatible systems and devices hindering progress and innovation.

Protocols, on the other hand, dictate the rules and procedures that govern data transmission and communication between devices. They outline the structure and format of messages, as well as the actions to be taken at each step of the communication process. Protocols ensure reliable and efficient communication by defining how data is packaged, transmitted, received, and interpreted.

This subchapter explores a wide range of telecommunications standards and protocols, including the most widely used ones such as TCP/IP, Ethernet, and Wi-Fi. It provides a detailed explanation of each standard or protocol, discussing their purpose, functionality, and key features. The content also covers emerging standards and protocols that are shaping the future of telecommunications engineering.

Moreover, the subchapter delves into the standardization bodies and organizations responsible for developing and maintaining these standards and protocols. It highlights the role of international bodies like the International Telecommunication Union (ITU) and the Institute of Electrical and Electronics Engineers (IEEE) in shaping the telecommunications landscape.

For electrical engineering professionals, this subchapter serves as a valuable resource, offering insights into the standards and protocols that underpin modern telecommunications. It provides a solid foundation for understanding the principles and concepts behind these standards, enabling engineers to design, develop, and troubleshoot complex telecommunications systems.

Whether you are a telecommunications engineer, a student, or someone interested in gaining a deeper understanding of the field, this subchapter empowers you with the knowledge required to navigate the intricate world of telecommunications standards and protocols.

OSI Model

In the ever-evolving world of telecommunications, understanding the fundamentals is crucial for electrical engineers and anyone interested in the field. One such fundamental concept is the OSI (Open Systems Interconnection) model. This subchapter aims to provide a comprehensive overview of the OSI model, its layers, and its significance in modern telecommunications.

The OSI model is a conceptual framework that standardizes the functions of a communication system into seven distinct layers. Each layer has a specific role to play, contributing to the overall efficiency and effectiveness of data transmission. By dividing the complex process of network communication into manageable layers, the OSI model simplifies troubleshooting and facilitates the interoperability of different systems.

The seven layers of the OSI model are as follows: physical, data link, network, transport, session, presentation, and application. The physical layer is responsible for transmitting raw bit streams over physical media, such as cables or wireless connections. The data link layer ensures error-free communication between adjacent devices, while the network layer handles logical addressing and routing. The transport layer ensures reliable end-to-end data delivery, while the session layer establishes and manages connections between applications. The presentation layer deals with data formatting and encryption, while the application layer provides services directly to the end user.

Understanding the OSI model is essential for electrical engineers, as it serves as a framework for designing and implementing communication protocols and network architectures. By adhering to the guidelines set by the OSI model, engineers can ensure compatibility and seamless integration of different systems, regardless of their manufacturers or origins. This standardized approach promotes interoperability, simplifies system design, and enhances overall network performance.

Moreover, the OSI model provides a foundation for troubleshooting and diagnosing network issues. By identifying the layer at which a problem occurs, engineers can focus their efforts on that specific layer, reducing the time and resources needed for resolution. Additionally, the OSI model allows for modular upgrades and modifications, as changes in one layer do not necessarily affect the others, ensuring scalability and future-proofing of communication systems.

In conclusion, the OSI model is a fundamental concept in telecommunications engineering. By understanding its layers and functions, electrical engineers can design, implement, and troubleshoot complex communication systems effectively. The OSI model's standardization and modularity contribute to the interoperability, scalability, and overall performance of modern networks. Whether you are a student, professional, or simply curious about the field of electrical engineering, a solid understanding of the OSI model is essential for success in the telecommunications industry.

TCP/IP Protocol Suite

The TCP/IP protocol suite is a foundational framework of protocols used for communication over the internet and other computer networks. This subchapter aims to provide a comprehensive understanding of the TCP/IP protocol suite, its components, and their functions. Whether you are an electrical engineering student, a professional in the field, or simply curious about the inner workings of the internet, this subchapter will be a valuable resource for you.

At its core, the TCP/IP protocol suite consists of two main protocols: the Transmission Control Protocol (TCP) and the Internet Protocol (IP). TCP is responsible for ensuring reliable transmission of data packets between devices by establishing and maintaining a connection. IP, on the other hand, handles the routing and addressing of packets, allowing them to be sent and received across different networks.

In addition to TCP and IP, the TCP/IP protocol suite encompasses several other protocols that work together to provide various functionalities. These protocols include the User Datagram Protocol (UDP), Internet Control Message Protocol (ICMP), Internet Group Management Protocol (IGMP), and many more. Each protocol has a specific role and contributes to the overall efficiency and effectiveness of the TCP/IP suite.

Understanding the TCP/IP protocol suite is essential for electrical engineers as it forms the backbone of modern communication networks. It enables devices to connect and communicate seamlessly, facilitating the exchange of information and enabling the internet as we know it today. By delving into the intricacies of the TCP/IP protocol suite, electrical engineers can gain insights into network design, troubleshooting, and optimization.

Moreover, this subchapter will explore the evolution of the TCP/IP protocol suite, from its origins in the early days of ARPANET to its widespread adoption and standardization. It will also discuss the different layers of the TCP/IP model, such as the network interface, internet, transport, and application layers, and how they interact with each other.

Whether you are a student eager to grasp the fundamentals of networking or a practicing electrical engineer seeking to enhance your knowledge, this subchapter will provide you with a comprehensive understanding of the TCP/IP protocol suite. With this knowledge, you will be better equipped to design, configure, and troubleshoot communication networks, making you an invaluable asset in the field of electrical engineering.

Wireless Communication Standards

In the rapidly evolving field of telecommunications engineering, wireless communication standards play a pivotal role in enabling seamless connectivity and efficient data transmission. These standards dictate the rules and protocols that govern wireless communication systems, ensuring compatibility and interoperability across various devices and networks. This subchapter will provide an overview of wireless communication standards, their significance, and their impact on the field of electrical engineering.

Wireless communication standards are a set of guidelines and specifications that define the technical aspects of wireless networks and devices. These standards are developed and maintained by international organizations such as the Institute of Electrical and Electronics Engineers (IEEE) and the International Telecommunication Union (ITU). They encompass a wide range of technologies, including cellular networks, Wi-Fi, Bluetooth, and satellite communication.

The importance of wireless communication standards cannot be overstated. They enable the seamless integration of different wireless devices and networks, allowing users to connect and communicate effortlessly. These standards ensure that devices from different manufacturers can communicate with each other, regardless of the underlying technology or frequency band being used. This interoperability promotes competition, innovation, and widespread adoption of wireless technologies.

For electrical engineers, understanding wireless communication standards is crucial. These standards provide the foundation for designing and implementing wireless systems, ensuring the efficient utilization of resources and the reliable transmission of data. Engineers working in the field of wireless communications must have a thorough knowledge of these standards to develop cutting-edge technologies and overcome the challenges associated with wireless network design, optimization, and security.

Moreover, the study of wireless communication standards offers numerous career opportunities for electrical engineers. With the ever-increasing demand for wireless connectivity, professionals skilled in implementing and maintaining wireless networks are in high demand. By mastering wireless communication standards, electrical engineers can position themselves as valuable assets in industries such as telecommunications, Internet of Things (IoT), and smart cities.

In this subchapter, we will delve into the fundamental wireless communication standards, such as IEEE 802.11 (Wi-Fi), IEEE 802.15 (Bluetooth), and cellular standards like 3G, 4G, and 5G. We will explore the technical aspects of these standards, including their frequency bands, modulation schemes, data rates, and security protocols. Additionally, we will discuss emerging wireless standards, such as Wi-Fi 6, 5G NR (New Radio), and the upcoming 6G, which promise even faster speeds, lower latency, and enhanced connectivity.

By the end of this subchapter, readers will have a comprehensive understanding of wireless communication standards, their significance in the field of electrical engineering, and the potential they hold for shaping the future of wireless connectivity. Whether you are an electrical engineering student, a seasoned professional, or simply curious about the fascinating world of wireless communications, this subchapter will provide you with the necessary knowledge to navigate the complex landscape of wireless communication standards.

Chapter 3: Telecommunication Networks

Introduction to Telecommunication Networks

Telecommunication networks are the backbone of our modern world, enabling us to communicate and connect with each other across vast distances. In this subchapter, we will delve into the fundamentals of telecommunication networks, exploring how they function and their significance in the field of electrical engineering.

Telecommunication networks are complex systems designed to transmit and receive information, such as voice, data, and video, between multiple users. These networks consist of various interconnected components, including switches, routers, transmission media, and end devices. Understanding the intricacies of these components is crucial for electrical engineers working in the telecommunications industry.

At the heart of any telecommunication network is the concept of connectivity. Networks are built to establish reliable connections between users, allowing them to exchange information seamlessly. This connectivity can be achieved through different technologies, such as wired connections using copper or fiber optic cables, or wireless connections through radio waves or satellite communications.

The subchapter will begin by introducing the basic concepts of telecommunication networks, including the different types of networks such as local area networks (LANs), wide area networks (WANs), and metropolitan area networks (MANs). We will discuss the advantages and disadvantages of each type, as well as their applications in various settings.

Next, we will explore the components that make up a telecommunication network. This will include an in-depth look at switches, routers, and transmission media, discussing their functions and how they contribute to the overall network infrastructure. We will also cover topics such as network protocols, addressing schemes, and data transmission methods.

Furthermore, the subchapter will address emerging trends and technologies in telecommunication networks. This will include discussions on virtual private networks (VPNs), cloud computing, Internet of Things (IoT), and the impact of 5G technology on network infrastructure.

Throughout the subchapter, we will provide practical examples and real-world applications to help illustrate the concepts discussed. Additionally, we will highlight the importance of telecommunication networks in our daily lives, from enabling seamless communication to supporting critical infrastructure such as healthcare and transportation systems.

Whether you are an electrical engineering student, a professional in the field, or simply someone interested in learning about telecommunication networks, this subchapter will serve as a comprehensive guide to mastering the fundamentals of this fascinating and ever-evolving domain.

Network Topologies

In the field of electrical engineering, network topologies play a vital role in the design and implementation of telecommunications systems. Whether it is a small local area network (LAN) or a vast global network, understanding different topologies is essential for telecommunication engineers. This subchapter aims to provide a comprehensive overview of network topologies, their types, advantages, and limitations.

A network topology refers to the physical or logical layout of interconnected devices in a network. It determines how data flows between devices and influences the network's overall performance and reliability. There are various types of network topologies, each with its unique characteristics.

One of the most common topologies is the bus topology. In this arrangement, all devices are connected to a single communication line, often referred to as a backbone. While it is cost-effective and easy to implement, a bus topology suffers from a single point of failure and limited scalability.

Another popular topology is the star topology, where all devices are connected to a central hub or switch. This configuration offers better reliability, easy troubleshooting, and scalability. However, it requires more cabling and can be expensive to implement.

A ring topology forms a closed loop, where each device is connected to its adjacent devices. Data flows in a circular manner, ensuring equal access to the network. However, a failure in one device can disrupt the entire network, making it less reliable.

Mesh topologies, on the other hand, provide redundancy and fault tolerance by connecting each device to every other device. This ensures alternative paths for data transmission, minimizing the risk of network downtime. However, mesh topologies are costly and complex to implement, making them suitable for critical applications.

Hybrid topologies combine two or more types of network topologies, allowing for greater flexibility and customization. This approach enables engineers to design networks that meet specific requirements, such as improved performance, fault tolerance, or cost-effectiveness.

Understanding the advantages and limitations of different network topologies is crucial for telecommunication engineers. Factors like network size, cost, performance requirements, and scalability need to be considered when selecting a topology for a specific application.

In conclusion, network topologies are fundamental building blocks of telecommunication systems. Electrical engineers must have a comprehensive understanding of different topologies to design and implement efficient and reliable networks. By carefully considering the advantages and limitations of each topology, engineers can create networks that meet the needs of various applications in the field of electrical engineering.

Wired Networks

Wired networks play a crucial role in the field of electrical engineering and are the backbone of modern telecommunication systems. In this subchapter, we will explore the fundamentals of wired networks, their components, and their importance in the field of telecommunications.

A wired network, also known as a physical network, is a communication network that uses physical cables to connect devices and transmit data. These cables can be made of copper, such as Ethernet cables, or fiber optic cables, which offer higher speeds and greater bandwidth. Wired networks provide a reliable and secure means of transferring data over long distances, making them indispensable in various industries.

One of the key components of a wired network is the network interface card (NIC), which enables devices to connect to the network. The NIC facilitates the transmission and reception of data packets between devices, allowing for seamless communication. Additionally, switches and routers are vital components that ensure efficient data transfer within the network.

Wired networks offer several advantages over wireless networks, particularly in terms of stability and security. Unlike wireless networks, which can be susceptible to interference and signal degradation, wired networks provide a consistent and reliable connection. This stability is especially crucial in critical applications, such as industrial automation or medical infrastructure, where downtime or data loss can have severe consequences.

Furthermore, wired networks offer enhanced security as the physical cables limit unauthorized access to the network. Wireless networks, on the other hand, can be vulnerable to unauthorized access, making them more susceptible to cyber-attacks. By utilizing wired networks, electrical engineers can design and implement robust security protocols to protect sensitive data and ensure the integrity of the network.

In the field of electrical engineering, understanding wired networks is essential for designing and implementing efficient telecommunication systems. From managing data centers to developing communication protocols, electrical engineers need to have a comprehensive understanding of wired network principles.

In conclusion, wired networks are the foundation of modern telecommunication systems and are of utmost importance in the field of electrical engineering. Their reliability, stability, and enhanced security make them an ideal choice for various applications. By mastering the principles of wired networks, electrical engineers can contribute to the development of cutting-edge telecommunication infrastructure and ensure the seamless transfer of data in today's connected world.

Local Area Networks (LANs)

In today's interconnected world, Local Area Networks (LANs) play a vital role in the field of electrical engineering. LANs are a fundamental component of modern telecommunications systems, enabling efficient and secure communication within a limited geographic area, such as an office building, university campus, or factory floor. This subchapter will provide a comprehensive overview of LANs, their components, and their applications, catering to a diverse audience interested in electrical engineering.

LANs are designed to connect multiple devices, such as computers, printers, servers, and switches, within a confined space, allowing them to share resources, exchange data, and collaborate effectively. The key advantage of LANs is their high data transfer rates, which facilitate quick and reliable communication between connected devices. This subchapter will explore the various types of LAN topologies, including bus, star, ring, and mesh, outlining their strengths and weaknesses.

Moreover, the subchapter will delve into LAN protocols and standards, such as Ethernet, Wi-Fi, and Token Ring, explaining their functionalities, differences, and current advancements. The content will highlight the importance of adhering to industry standards to ensure compatibility and interoperability between different LAN devices and technologies.

Security is another critical aspect of LANs, as they often handle sensitive and confidential information. This subchapter will discuss common security measures, including firewalls, encryption, access control, and intrusion detection systems, providing readers with a comprehensive understanding of how to protect LANs from unauthorized access and data breaches.

Furthermore, the subchapter will explore the practical applications of LANs in various sectors, such as business, education, healthcare, and manufacturing. It will showcase real-world examples of LAN implementations, including the design considerations, challenges faced, and benefits achieved.

Throughout the subchapter, practical tips, best practices, and case studies will be provided to deepen the reader's understanding and encourage the application of LAN concepts in their professional endeavors. Whether a student, a novice engineer, or an experienced professional, this subchapter will serve as a valuable resource for anyone interested in mastering the intricacies of Local Area Networks (LANs) in the field of electrical engineering.

Metropolitan Area Networks (MANs)

In the world of telecommunications engineering, Metropolitan Area Networks (MANs) play a crucial role in connecting various local area networks (LANs) within a metropolitan area. A MAN serves as a bridge between LANs, enabling efficient and reliable data transmission across a larger geographical area. This subchapter will delve into the intricacies of MANs, offering a comprehensive understanding of their architecture, components, and applications.

A Metropolitan Area Network is a high-speed network that spans a city or a metropolitan region, typically covering an area of up to 100 kilometers. MANs are designed to support a large number of users and facilitate the transmission of vast amounts of data. They are commonly used by organizations such as universities, government agencies, and large corporations to interconnect their various branches or campuses.

The architecture of a MAN usually involves a combination of fiber optic cables, microwave links, and leased lines. These transmission media ensure high bandwidth and low latency, allowing for the seamless exchange of data between different LANs. MANs are often built using a hierarchical structure, with a core network at the center, connecting to distribution networks, which, in turn, connect to the end-user LANs.

The components of a MAN include routers, switches, and multiplexers. Routers are responsible for forwarding data packets between different networks, ensuring that they reach their intended destinations. Switches, on the other hand, connect devices within a LAN, enabling efficient data transfer. Multiplexers are used to combine multiple data streams into a single stream, optimizing bandwidth usage.

The applications of MANs are diverse and far-reaching. They allow for the seamless integration of various services, such as voice, video, and data, over a single network infrastructure. MANs also support the implementation of advanced technologies like cloud computing, which require high-speed and reliable connectivity. Additionally, MANs facilitate the creation of virtual private networks (VPNs), enabling secure communication between geographically dispersed entities.

In conclusion, Metropolitan Area Networks (MANs) form an essential part of the telecommunications infrastructure within metropolitan areas. Their architecture, components, and applications make them a vital tool for organizations in the field of electrical engineering. Understanding the intricacies of MANs is crucial for telecom engineers, as they continue to play a pivotal role in the development of efficient and reliable communication networks.

Wide Area Networks (WANs)

Wide Area Networks (WANs) are a critical component of modern telecommunications systems. In this subchapter, we will explore the fundamentals of WANs, their technologies, and their significance in the field of electrical engineering.

A Wide Area Network (WAN) refers to a network that spans large geographical areas, connecting multiple local area networks (LANs) across different locations. WANs enable organizations to establish seamless communication and data sharing capabilities between remote sites, branches, or even countries. They play a pivotal role in connecting people and businesses globally, facilitating the exchange of information and promoting collaboration.

One of the key technologies employed in WANs is the use of dedicated communication lines, such as leased lines or fiber optic cables. These lines provide high bandwidth and low latency, ensuring efficient and reliable data transmission over long distances. Additionally, WANs can utilize various networking protocols, such as Multiprotocol Label Switching (MPLS) or Asynchronous Transfer Mode (ATM), to prioritize and manage the flow of data traffic, optimizing network performance.

WANs offer numerous advantages for electrical engineering professionals. They enable remote monitoring and control of electrical systems, allowing engineers to access and manage critical infrastructure from a central location. This capability is particularly valuable for large-scale power grids or industrial automation systems. Moreover, WANs facilitate the seamless integration of renewable energy sources, enabling engineers to monitor and control distributed power generation resources over vast territories.

As the demand for data-intensive applications continues to rise, WANs also play a crucial role in providing high-speed internet connectivity to users. They enable efficient video conferencing, cloud-based services, and remote access to resources, fostering collaboration and productivity in the field of electrical engineering. Additionally, WANs facilitate the implementation of advanced technologies like the Internet of Things (IoT) and Artificial Intelligence (AI) in electrical systems, paving the way for smart grid solutions and efficient energy management.

In conclusion, Wide Area Networks (WANs) are an essential component of modern telecommunications and electrical engineering. They enable seamless communication and data sharing between geographically dispersed locations, leveraging dedicated communication lines and networking protocols. WANs empower electrical engineers to monitor and control critical infrastructure remotely, integrate renewable energy sources, and embrace emerging technologies. As the world becomes increasingly interconnected, understanding and mastering the intricacies of WANs is crucial for professionals in the field of electrical engineering.

Wireless Networks

In today's interconnected world, wireless networks have become an essential part of our daily lives. From smartphones and laptops to smart home devices and industrial applications, wireless communication has revolutionized the way we connect and communicate. This subchapter will provide a comprehensive overview of wireless networks, exploring the fundamental concepts, technologies, and applications that are shaping the field of electrical engineering.

Introduction to Wireless Networks: Wireless networks refer to the communication networks that utilize wireless signals to transmit information. Unlike traditional wired networks that rely on physical cables, wireless networks enable devices to connect and communicate through radio waves, infrared signals, or satellite links. This flexibility and mobility have contributed to the widespread adoption and popularity of wireless networks across various industries.

Wireless Network Technologies: There are several wireless network technologies, each designed for specific applications and environments. This subchapter will delve into the most common wireless technologies, including Wi-Fi, Bluetooth, cellular networks, satellite communications, and RFID (Radio Frequency Identification). By understanding the characteristics and capabilities of these wireless technologies, electrical engineering professionals can effectively design and implement wireless networks to meet specific requirements.

Wireless Network Architecture:
To establish a wireless network, it is crucial to understand its architecture. This subchapter will explore the key components of wireless network architecture, such as access points, network controllers, and client devices. Additionally, it will cover concepts like network topologies, frequency bands, and signal propagation, providing a holistic understanding of the infrastructure necessary to create robust wireless networks.

Wireless Network Security:
With the increasing reliance on wireless networks, ensuring their security has become a critical concern. This subchapter will address various security measures and protocols used to protect wireless networks from unauthorized access, data breaches, and cyber threats. Topics such as encryption, authentication, and intrusion detection systems will be discussed, equipping electrical engineering professionals with the knowledge to design secure wireless networks.

Applications of Wireless Networks:
The applications of wireless networks are vast and diverse. This subchapter will explore the wide-ranging domains where wireless networks are employed, including telecommunications, healthcare, transportation, smart cities, and the Internet of Things (IoT). By understanding the specific requirements and challenges of each application, electrical engineering professionals can tailor wireless networks to meet the unique needs of their respective industries.

Conclusion:
Wireless networks have revolutionized the way we connect and communicate, offering flexibility, mobility, and convenience. This

subchapter has provided a comprehensive overview of wireless networks, covering the fundamental concepts, technologies, architectures, security measures, and applications. By mastering the principles of wireless networks, electrical engineering professionals can contribute to the advancement of this ever-evolving field and harness its potential to shape the future of telecommunications.

Cellular Networks

Cellular networks have become an integral part of our everyday lives, providing us with seamless connectivity and enabling us to communicate with others no matter where we are. In this subchapter, we will delve into the world of cellular networks, exploring their architecture, technologies, and the impact they have had on the field of electrical engineering.

Cellular networks are a type of wireless network that operates through a system of interconnected base stations covering a specific geographical area known as a cell. These cells are designed to overlap, ensuring continuous coverage and allowing for seamless handoff as a user moves from one cell to another. The architecture of a cellular network consists of several key components, including the base stations, mobile switching centers, and the core network. Each component plays a crucial role in ensuring the smooth functioning of the network.

One of the most significant advancements in cellular networks is the evolution of different generations, from 1G to the current 5G technology. Each generation has brought about significant improvements in terms of data transfer speeds, capacity, and reliability. 5G, the latest generation, promises ultra-low latency and blazing fast download speeds, making it a game-changer in various fields such as autonomous vehicles, remote surgery, and the Internet of Things (IoT).

The development of cellular networks has been a driving force behind many technological breakthroughs in the field of electrical engineering. Engineers have had to design and optimize antenna systems, develop efficient modulation schemes, and improve signal processing algorithms to ensure reliable and high-quality communication over cellular networks. Additionally, the increasing demand for data has led to the development of advanced technologies such as Multiple Input Multiple Output (MIMO) and beamforming, which have revolutionized wireless communication.

Furthermore, the advent of cellular networks has opened up a world of possibilities for electrical engineers. It has created numerous job opportunities, ranging from designing and deploying network infrastructure to developing innovative applications and services that leverage the power of cellular connectivity. As the demand for connectivity continues to grow, electrical engineers will play a crucial role in expanding and improving cellular networks to meet the needs of a digital society.

In conclusion, cellular networks have transformed the way we communicate, enabling us to stay connected wherever we go. Their architecture, technologies, and impact on electrical engineering make them a fascinating subject to explore. Whether you are an electrical engineering student, a professional in the field, or simply curious about the technology that powers your smartphone, this subchapter will provide you with valuable insights into the world of cellular networks.

Satellite Communication

Satellite communication is a crucial pillar of modern telecommunications engineering. It plays a vital role in connecting people and providing seamless communication services across vast distances. In this subchapter, we will explore the fundamentals of satellite communication, its components, and its applications in various fields.

Satellite communication involves the transmission of signals between a ground station and a satellite in orbit. The satellite acts as a relay station, receiving signals from one location and transmitting them to another. This technology enables global communication coverage, making it possible to connect people in remote areas, maritime vessels, airplanes, and even in outer space.

The key components of a satellite communication system include the satellite itself, ground stations for transmitting and receiving signals, and the communication links that connect them. Satellites can be classified into different types, such as geostationary satellites, low-Earth orbit satellites, and medium-Earth orbit satellites, each with its own advantages and applications.

Geostationary satellites are positioned at an altitude of approximately 36,000 kilometers above the Earth's equator, allowing them to remain in a fixed position relative to the Earth's surface. This characteristic enables uninterrupted communication coverage over a specific geographic area. On the other hand, low-Earth orbit and medium-Earth orbit satellites are positioned at lower altitudes, providing advantages such as lower latency and higher data rates.

Satellite communication finds applications in various fields, including broadcasting, telecommunications, navigation systems, and weather forecasting. It enables the transmission of television and radio signals to a wide audience, ensuring reliable and widespread access to news, entertainment, and educational content. Additionally, satellite communication is essential for global positioning systems (GPS) used in navigation and tracking services.

In the field of telecommunications, satellite communication provides a reliable backup for terrestrial networks, ensuring uninterrupted communication during natural disasters or emergencies. It also facilitates long-distance phone calls, internet connectivity, and broadband services in remote areas where the installation of traditional infrastructure is challenging.

In conclusion, satellite communication is a crucial technology in the field of telecommunications engineering. It enables global connectivity, provides essential services in various sectors, and ensures reliable communication even in remote locations. Understanding the fundamentals of satellite communication is vital for electrical engineering professionals and anyone interested in the field of telecommunications.

Wireless Local Area Networks (WLANs)

In today's fast-paced world, where connectivity is an integral part of our daily lives, Wireless Local Area Networks (WLANs) have become increasingly essential. This subchapter aims to provide a comprehensive understanding of WLANs, their applications, and the underlying principles that make them possible. Whether you are an electrical engineering student, a professional in the field, or simply an individual interested in telecommunications, this subchapter will serve as a valuable resource.

WLANs, also known as Wi-Fi networks, enable wireless communication between devices within a limited geographical area. They have revolutionized the way we access information, allowing us to connect our devices seamlessly to the internet without the constraints of physical cables. From homes and offices to public spaces such as cafes and airports, WLANs have become ubiquitous, providing us with the freedom to access information and communicate on the go.

This subchapter will delve into the technical aspects of WLANs, starting with the fundamentals of radio frequency (RF) communication. Understanding the basics of RF signals, modulation, and transmission will lay the foundation for comprehending WLAN principles. We will explore the different WLAN standards, such as IEEE 802.11a/b/g/n/ac, and their respective capabilities, frequencies, and data rates. Additionally, we will discuss the challenges associated with RF interference, security, and the mitigation techniques used to address them.

Furthermore, this subchapter will cover WLAN architecture, including the components that make up a WLAN system. We will examine the roles of access points (APs), wireless clients, and the importance of radio frequency planning to ensure optimal coverage and performance. We will also delve into WLAN deployment options, such as centralized and distributed architectures, and the factors to consider when designing a WLAN network.

To provide a well-rounded understanding of WLANs, we will explore emerging technologies and trends that are shaping the future of wireless communication. Topics such as Wi-Fi 6 (802.11ax), mesh networks, and the integration of WLANs with other wireless technologies will be discussed, giving you insights into the latest advancements in the field.

Whether you are a student, professional, or simply curious about the world of electrical engineering, this subchapter on Wireless Local Area Networks (WLANs) will equip you with the knowledge needed to understand, design, and deploy WLAN systems effectively. Stay connected, explore the possibilities, and embrace the wireless revolution.

Chapter 4: Telecommunication Systems

Telephone Systems

Telephone systems have revolutionized communication, connecting people across vast distances and allowing for efficient and effective conversations. This subchapter explores the intricacies of telephone systems, delving into the various components and technologies that make them work. Aimed at electrical engineering professionals and enthusiasts, this comprehensive guide provides a comprehensive overview of telephone systems.

Introduction to Telephone Systems

Telephone systems serve as the backbone of modern communication networks, enabling voice conversations, data transmission, and internet access. This subchapter delves into the fundamental concepts and technologies that underpin telephone systems, offering readers a solid foundation to understand the complexities of this field.

Components of Telephone Systems

To grasp the functioning of telephone systems, it is essential to understand the key components involved. This subchapter explores the different elements that make up telephone systems, including telephones, exchange switches, transmission facilities, and subscriber lines. Readers will gain insights into how these components work together to facilitate communication.

Digital Telephony

The advent of digital technology has significantly transformed telephone systems. This subchapter delves into the principles and advantages of digital telephony, exploring concepts such as pulse code modulation (PCM), time-division multiplexing (TDM), and the use of digital signals for voice transmission. An in-depth understanding of digital telephony is crucial for electrical engineering professionals working in the telecommunications industry.

Telephone Switching Systems

Telephone switching systems play a pivotal role in establishing and maintaining connections between callers. This subchapter discusses various switching technologies, including circuit-switching, packet-switching, and hybrid switching. It explains the functionalities of different types of switches and their role in ensuring smooth communication flow.

Emerging Technologies in Telephone Systems

As technology continues to evolve, so do telephone systems. This subchapter sheds light on emerging technologies that are shaping the future of telecommunications. Topics covered include Voice over Internet Protocol (VoIP), mobile telephony, and wireless communication. Electrical engineering professionals will gain insights into the latest advancements and trends in telephone systems.

Conclusion

Telephone systems have evolved significantly over the years, revolutionizing the way we communicate. This subchapter has provided a comprehensive overview of telephone systems, covering the fundamental components, digital telephony, switching systems, and emerging technologies. By mastering the concepts and technologies discussed, electrical engineering professionals will be well-equipped to contribute to the ever-evolving field of telecommunications.

Analog Telephone Systems

Analog telephone systems have played a significant role in the development of telecommunications engineering. Before the advent of digital communication technologies, analog systems were the primary means of transmitting voice signals over long distances. This subchapter will delve into the fundamentals of analog telephone systems, exploring their operation, components, and limitations.

An analog telephone system converts voice signals into electrical signals that can be transmitted over copper wires. The system comprises various components, including telephones, exchange switches, and transmission lines. When a user speaks into a telephone, their voice is converted into electrical signals by a microphone. These signals are then amplified and modulated onto a carrier frequency for transmission.

Exchange switches serve as the backbone of analog telephone systems. They act as the central hub for connecting calls between different subscribers. Exchange switches receive signals from the calling subscriber, identify the destination number, and establish a connection with the receiving subscriber. This process involves several steps, including digitizing the analog voice signals, routing the call, and demodulating the signals at the receiving end.

One of the key advantages of analog telephone systems is their simplicity. The technology is relatively easy to understand and implement, making it accessible to electrical engineering professionals. Additionally, analog systems have proven to be reliable and robust, with the ability to operate during power outages.

However, analog telephone systems do have their limitations. One major constraint is the limited bandwidth available for transmitting voice signals. Analog systems can only transmit a narrow range of frequencies, resulting in reduced audio quality compared to digital systems. Moreover, analog systems are susceptible to noise and interference, which can degrade the signal quality further.

As digital telecommunications technologies continue to advance, analog telephone systems are gradually being replaced by digital systems. However, analog systems still have a presence in some areas, particularly in rural regions with limited infrastructure. Understanding the principles and operation of analog telephone systems is crucial for electrical engineering professionals, as it provides a foundational knowledge of telecommunication systems.

In conclusion, analog telephone systems have played a vital role in the history of telecommunications engineering. This subchapter has provided an overview of the operation, components, and limitations of analog systems. While digital technologies have largely superseded analog systems, understanding the fundamentals of analog telephony remains relevant for electrical engineering professionals working in the field of telecommunications.

Digital Telephone Systems

In today's rapidly advancing technological landscape, digital telephone systems have become a cornerstone of modern communication networks. This subchapter will provide a comprehensive overview of digital telephone systems, delving into their fundamental principles, benefits, and the various technologies that underpin their functionality.

Digital telephone systems utilize digital signals to transmit voice and data over telecommunications networks. Unlike traditional analog systems, which transmit signals in continuous waveforms, digital systems convert voice and data into binary code, allowing for more efficient transmission and enhanced clarity.

One of the significant advantages of digital telephone systems is their ability to compress voice signals, resulting in higher call capacity within the same bandwidth. This compression technique, known as Pulse Code Modulation (PCM), converts analog voice signals into a digital format by sampling the signal at regular intervals. PCM revolutionized the telecommunications industry by enabling the transmission of multiple voice channels simultaneously over a single connection.

Another key feature of digital telephone systems is their integration with the Internet Protocol (IP). This convergence of voice and data networks has given rise to Voice over IP (VoIP) technology. VoIP allows users to make calls over the internet, eliminating the need for traditional telephone lines and reducing costs significantly. Additionally, VoIP offers a range of advanced features such as video conferencing, call forwarding, and voicemail to email transcription, enhancing the overall communication experience.

Digital telephone systems also incorporate various signaling protocols for call setup, management, and control. The most widely used protocol is the Signaling System 7 (SS7), which enables seamless communication between different telephone networks and facilitates advanced features such as caller ID and call waiting.

Moreover, advancements in digital telephone systems have led to the development of Unified Communications (UC) solutions. UC integrates various communication channels, including voice, video, instant messaging, and email, into a single platform, enhancing productivity and collaboration across organizations.

In conclusion, digital telephone systems have revolutionized the telecommunications industry, providing enhanced call quality, increased call capacity, and a wide range of advanced features. From residential to enterprise environments, digital telephone systems have become an indispensable tool for efficient and effective communication. By embracing these technologies, electrical engineering professionals can gain a comprehensive understanding of the underlying principles and technologies driving digital telephone systems, enabling them to contribute to the continued evolution and advancement of this vital field.

Voice over IP (VoIP)

Voice over IP (VoIP) technology has revolutionized the field of telecommunications, offering a cost-effective and efficient way to transmit voice and multimedia content over the internet. In this subchapter, we will explore the intricacies of VoIP and its importance in the realm of electrical engineering.

VoIP refers to the transmission of voice and multimedia content over IP networks, such as the internet. Unlike traditional telephony systems that rely on dedicated copper lines, VoIP utilizes packet-switched networks to transmit data in small packets. These packets are then reassembled at the receiving end to recreate the original voice or multimedia content.

The adoption of VoIP has been driven by its numerous advantages. Firstly, it offers significant cost savings compared to traditional telephony systems. By using the existing internet infrastructure, VoIP eliminates the need for separate voice and data networks, reducing maintenance and operational costs.

Furthermore, VoIP provides improved flexibility and scalability. With traditional phone systems, adding or removing phone lines can be a cumbersome and expensive process. In contrast, VoIP allows for easy scalability, enabling businesses to add or remove lines as per their requirements.

Another key advantage of VoIP is its ability to integrate various communication services. VoIP systems can seamlessly integrate voice, video, and data services, enabling users to make video calls, share files, and conduct conferences, all through a single platform. This integration enhances productivity and collaboration within organizations.

From an electrical engineering perspective, VoIP involves complex protocols and technologies. Engineers need to understand the underlying principles of packet-switched networks, compression algorithms, and quality of service (QoS) mechanisms. They also need to be proficient in network design and troubleshooting to ensure optimal performance and reliability of VoIP systems.

In conclusion, VoIP is a transformative technology in the field of telecommunications, providing cost savings, flexibility, and enhanced communication services. Electrical engineers play a crucial role in designing, implementing, and maintaining VoIP systems. By leveraging their expertise in network protocols and technologies, electrical engineers contribute to the seamless functioning of VoIP networks, enabling individuals and organizations to communicate effectively and efficiently.

Data Communication Systems

Data communication systems play a vital role in today's interconnected world. From simple email exchanges to complex global networks, these systems are the backbone of modern communication. This subchapter will delve into the fundamental concepts and technologies behind data communication systems, offering a comprehensive guide for individuals interested in electrical engineering and its related fields.

The subchapter will begin by introducing the basic principles of data communication, emphasizing the importance of data transmission and the need for reliable and efficient communication systems. It will discuss the various types of data that can be transmitted, such as text, images, audio, and video, and highlight the challenges associated with each.

Next, the subchapter will explore the different components of a data communication system. It will cover topics such as data sources, transmission media, and data receivers, explaining how these elements work together to facilitate the transfer of information. The subchapter will also address the concept of data encoding and decoding, shedding light on techniques used to convert data into a suitable format for transmission.

Furthermore, the subchapter will delve into the various types of data communication networks. It will discuss local area networks (LANs), wide area networks (WANs), and metropolitan area networks (MANs), providing insights into their respective architectures, topologies, and protocols. The subchapter will also touch upon emerging technologies like wireless communication systems and the Internet of Things (IoT), highlighting their impact on data communication.

To ensure a comprehensive understanding, the subchapter will feature real-life examples, case studies, and practical applications of data communication systems. It will discuss the challenges faced in designing and implementing these systems, along with strategies to overcome them. Additionally, the subchapter will address the importance of network security and data protection, emphasizing the need for robust measures to safeguard sensitive information.

Overall, this subchapter on data communication systems aims to equip readers with a solid foundation in the field of electrical engineering. Whether you are a student, a professional, or simply curious about the inner workings of modern communication, this subchapter will provide valuable insights into the fascinating world of data communication systems.

Modems and Data Transmission

In today's interconnected world, where information is transmitted at lightning speed, understanding the fundamentals of modems and data transmission is crucial for both electrical engineering professionals and anyone interested in the field of telecommunications. This subchapter aims to provide a comprehensive guide to modems and data transmission, covering the basic principles, technologies, and applications involved.

Modems, short for modulator-demodulators, are electronic devices that enable the transmission of data over communication channels, such as telephone lines or fiber-optic cables. They serve as the bridge between digital devices, such as computers, and the analog communication channels used for transmission. By converting digital signals into analog form for transmission and re-converting them back to digital at the receiving end, modems allow for reliable and efficient data transfer.

The subchapter starts by delving into the fundamentals of data transmission, explaining concepts such as data rate, bandwidth, and protocols. It explores the differences between analog and digital signals and how modems facilitate the conversion between them. Various modulation techniques, such as amplitude modulation (AM), frequency modulation (FM), and phase-shift keying (PSK), are discussed in detail, with their advantages and limitations highlighted.

The subchapter then moves on to explore the different types of modems available, including dial-up modems, cable modems, and digital subscriber line (DSL) modems. Each type is explained in terms of their working principles, data transfer speeds, and typical applications. The subchapter also explores the latest advancements in modem technology, such as wireless modems and satellite modems, and their implications for modern telecommunications.

Additionally, the subchapter addresses the challenges and considerations involved in modem design and implementation. Factors such as noise, signal interference, and error correction techniques are discussed, along with the importance of adopting robust modulation schemes to ensure reliable data transmission.

Overall, this subchapter serves as a comprehensive guide to modems and data transmission, providing valuable insights for electrical engineering professionals and anyone interested in understanding the intricacies of modern telecommunications. It equips the audience with the knowledge required to make informed decisions regarding modem selection, deployment, and optimization, ultimately contributing to the advancement of the field of electrical engineering in the ever-evolving world of telecommunications.

Local Area Networks (LANs)

In today's interconnected world, the need for efficient and reliable communication networks is paramount. Local Area Networks (LANs) have emerged as the foundation of modern telecommunication systems, enabling seamless data transfer and collaboration within a limited geographical area. This subchapter aims to provide a comprehensive overview of LANs, their components, and their role in the field of electrical engineering.

LANs are private networks that connect computers, devices, and peripherals within a small area, such as an office building, school, or home. They facilitate the sharing of resources, including files, printers, and internet connections, among connected devices. LANs utilize various technologies, including Ethernet, Wi-Fi, and Token Ring, to establish connections and transmit data.

Ethernet is the most widely used LAN technology, providing fast and reliable communication. It employs a set of protocols to enable devices to connect seamlessly and transmit data packets. Wi-Fi, on the other hand, enables wireless LAN connectivity, allowing devices to connect without physical cables. This technology has revolutionized the way we access the internet, providing flexibility and mobility to users.

LANs consist of several components that work harmoniously to facilitate data transfer. These components include network interface cards (NICs), routers, switches, hubs, and cables. NICs are installed in devices to enable network connectivity. Routers act as traffic managers, directing data packets between different networks. Switches, on the other hand, connect devices within the LAN and facilitate efficient data transfer. Hubs are simpler devices that connect multiple devices in a LAN, but they offer less functionality compared to switches.

Electrical engineers play a crucial role in designing and implementing LANs. They are responsible for analyzing network requirements, selecting appropriate technologies, and configuring network devices. They ensure that LANs are secured, scalable, and efficient, meeting the needs of the users. Electrical engineers also troubleshoot network issues and optimize LAN performance.

In conclusion, LANs are the backbone of modern telecommunication systems, enabling efficient communication and data transfer within a limited geographical area. Electrical engineers play a vital role in designing and implementing LANs, ensuring their seamless operation. Understanding LAN technologies and their components is essential for anyone interested in the field of electrical engineering, as LANs continue to evolve and shape our interconnected world.

Wide Area Networks (WANs)

In the realm of electrical engineering, the concept of Wide Area Networks (WANs) holds immense significance. WANs refer to a network infrastructure that spans a large geographical area, connecting multiple local area networks (LANs) or even individual devices across cities, countries, or continents. This subchapter delves into the intricacies of WANs, shedding light on their architecture, components, and the significance they hold in modern telecommunications.

WANs are instrumental in facilitating efficient communication and data transfer between geographically dispersed locations. They enable organizations to connect remote offices, data centers, and branch locations together, creating a seamless network environment. With the advent of cloud computing and the growing need for real-time data sharing, WANs have become the backbone of modern-day businesses and industries.

One of the key components of a WAN is the use of dedicated leased lines or public communication links to establish connectivity between different locations. These links, such as T1 lines or optical fiber cables, provide high-speed and reliable data transmission. Additionally, WANs often employ various networking technologies like Multiprotocol Label Switching (MPLS) or Virtual Private Networks (VPNs) to ensure secure and efficient data transfer.

The subchapter also explores the different types of WAN topologies, including point-to-point, star, mesh, and hybrid. Each topology offers unique advantages and is suited for specific use cases. Point-to-point connections, for instance, provide a direct link between two locations, ideal for small-scale networks. On the other hand, mesh topologies offer redundancy and reliability by establishing multiple connections between various nodes.

Moreover, the subchapter delves into the challenges faced in WAN design and management. Bandwidth limitations, latency, security concerns, and network congestion are some of the common obstacles that engineers encounter when designing WANs. Therefore, it is essential for electrical engineers to possess a comprehensive understanding of WAN principles to overcome these challenges.

In conclusion, Wide Area Networks (WANs) play a vital role in today's interconnected world. This subchapter serves as a comprehensive guide for electrical engineering professionals to grasp the fundamental concepts, architectures, and challenges associated with WANs. By mastering the intricacies of WANs, engineers will be better equipped to design, implement, and manage efficient network infrastructures that support seamless communication and data transfer across vast distances.

Wireless Communication Systems

Wireless communication systems have revolutionized the way we transmit and receive information, enabling seamless connectivity and communication across various devices and platforms. This subchapter explores the fundamental principles and technologies behind wireless communication, providing a comprehensive understanding of this rapidly evolving field.

Introduction to Wireless Communication Systems

In today's interconnected world, wireless communication systems play a crucial role in almost every aspect of our lives. Whether it's sending a text message, making a phone call, or streaming a video, wireless technologies enable us to communicate effortlessly, wirelessly, and in real-time.

This subchapter begins by introducing the basics of wireless communication systems, including the historical context and the key driving factors behind their development. It highlights the importance of wireless technology in bridging the digital divide and facilitating global connectivity.

Wireless Communication Technologies

Next, we delve into the various wireless communication technologies that have shaped the modern world. We explore the fundamentals of radio frequency (RF) and microwave communication, including modulation techniques, channel coding, and multiplexing schemes. The subchapter also discusses the evolution of wireless standards, such as 3G, 4G, and the current state-of-the-art 5G networks.

Wireless Network Architectures

To understand how wireless communication systems function, it is essential to grasp the underlying network architectures. This subchapter provides an overview of wireless network topologies, including cellular networks, ad hoc networks, and sensor networks. It explains the concept of base stations, access points, and wireless routers, highlighting their roles in enabling wireless connectivity.

Challenges and Future Trends

Wireless communication systems face numerous challenges, including limited bandwidth, interference, security threats, and spectrum allocation. This subchapter addresses these challenges and discusses the solutions and advancements made in each area. It also explores emerging trends, such as Internet of Things (IoT), smart cities, and wireless sensor networks, which promise to reshape the way we interact with the world.

Conclusion

Wireless communication systems have transformed the way we communicate and connect with each other. This subchapter provides a comprehensive overview of the principles, technologies, and network architectures that underpin wireless communication. Whether you are an electrical engineering student, a professional in the field, or simply someone interested in understanding the mechanics of wireless communication, this subchapter will equip you with the knowledge needed to navigate the wireless world. Stay tuned for the upcoming chapters, where we will delve deeper into specific wireless technologies, applications, and emerging trends.

Cellular Communication Systems

In today's interconnected world, cellular communication systems have become an integral part of our daily lives. From making phone calls to accessing the internet, these systems play a crucial role in keeping us connected. This subchapter aims to provide a comprehensive overview of cellular communication systems, their working principles, and their significance in the field of electrical engineering.

Cellular communication systems are networks of interconnected cells that allow for wireless communication between mobile devices and infrastructure. These systems are made up of various components, including base stations, mobile devices, and a central switching office. The base stations act as the link between the mobile devices and the network, providing coverage and facilitating communication.

The primary technology behind cellular communication systems is known as multiple access. This technology enables multiple users to share the available frequency spectrum, allowing for efficient and simultaneous communication. Two widely used multiple access techniques are Time Division Multiple Access (TDMA) and Code Division Multiple Access (CDMA). TDMA assigns time slots to different users, while CDMA assigns unique codes to each user.

The significance of cellular communication systems in the field of electrical engineering cannot be overstated. These systems have revolutionized the way we communicate, enabling wireless voice and data transmission over vast distances. The development of cellular networks has paved the way for advancements such as 4G and 5G technology, which offer faster data speeds and lower latency.

Moreover, cellular communication systems have also played a vital role in bridging the digital divide by providing connectivity to remote and underserved areas. They have empowered individuals and communities by giving them access to information, education, and economic opportunities.

As an electrical engineer, understanding the principles and working of cellular communication systems is crucial. This knowledge allows engineers to design and optimize these networks, ensuring efficient coverage and capacity. It also enables them to develop innovative solutions to challenges faced by these systems, such as interference and spectrum management.

In conclusion, cellular communication systems are an essential aspect of modern life and have greatly impacted the field of electrical engineering. This subchapter has provided a comprehensive overview of these systems, their working principles, and their significance. Whether you are a student, a professional, or simply curious about the technology that keeps us connected, this knowledge will undoubtedly enhance your understanding of the world of telecommunications.

Wireless Local Area Networks (WLANs)

In today's interconnected world, wireless technology has become an integral part of our lives. Wireless Local Area Networks (WLANs) have emerged as a key component of modern communication systems, providing seamless connectivity and enabling the exchange of information within a limited geographical area. This subchapter explores the fundamental concepts of WLANs, shedding light on their applications, standards, security, and future prospects.

WLANs are primarily used to establish wireless connectivity within a specific area such as homes, offices, or public spaces. They utilize radio frequency signals to transmit and receive data between devices, eliminating the need for physical cables. By leveraging the IEEE 802.11 family of standards, WLANs ensure interoperability among different devices and allow users to access the internet, share files, and communicate with each other effortlessly.

This subchapter dives into the technical aspects of WLANs, providing insights into the different standards that have evolved over time, including 802.11a, 802.11b, 802.11g, 802.11n, and the latest 802.11ac. It discusses the advantages and limitations of each standard, their data transfer rates, and frequency bands they operate on.

Security is a crucial concern in wireless networks, and this subchapter explores the various security mechanisms employed in WLANs. It delves into encryption protocols such as WEP, WPA, and WPA2, and highlights the importance of implementing strong passwords, MAC filtering, and network segmentation to safeguard against unauthorized access and data breaches.

Furthermore, this subchapter delves into the emerging trends and future prospects of WLANs. With the advent of Internet of Things (IoT) and the increasing demand for smart devices, WLANs are evolving to support a larger number of connected devices and higher data rates. The subchapter explores the challenges faced by WLANs in managing these increasing demands and the potential solutions being developed to address them, including the implementation of 5G technology and the use of advanced wireless technologies like beamforming and MU-MIMO.

Whether you are an electrical engineering student, a professional in the telecommunications field, or simply someone interested in understanding the workings of wireless networks, this subchapter provides a comprehensive overview of WLANs. By mastering the concepts presented here, you will be equipped with the knowledge to design, deploy, and secure wireless networks, contributing to the advancement of modern communication systems.

Satellite Communication Systems

Satellite communication systems have revolutionized the field of telecommunications, enabling global connectivity and seamless communication across vast distances. This subchapter aims to provide an overview of satellite communication systems, exploring their fundamental principles, components, and applications.

In simplest terms, a satellite communication system involves the transmission of signals from Earth to a satellite in space, which then relays the signals back to another location on Earth. These systems rely on the use of geostationary satellites, which orbit the Earth at the same rotational speed, allowing them to remain fixed relative to a specific location on Earth's surface. This unique characteristic makes them ideal for providing continuous coverage to a wide area.

One of the key components of a satellite communication system is the ground segment, which consists of various Earth-based stations responsible for transmitting and receiving signals to and from the satellites. These stations include antennas, transmitters, receivers, and other associated equipment. The satellite itself is equipped with similar components, including transponders that receive, amplify, and retransmit signals back to Earth.

Satellite communication systems find applications in various fields, including telephony, television broadcasting, internet connectivity, and remote sensing. They have played a significant role in bridging the digital divide by providing connectivity to remote and underserved areas, allowing people to access information, education, and healthcare services.

In the field of electrical engineering, a thorough understanding of satellite communication systems is essential. Electrical engineers are responsible for designing, developing, and maintaining the ground segment infrastructure, including the antennas, transmitters, and receivers. They also play a crucial role in optimizing the performance of satellite communication systems, ensuring efficient signal transmission, and minimizing interference.

Moreover, electrical engineers are involved in the design and development of satellite payloads, which are responsible for processing and relaying the signals. These payloads include advanced signal processing units, power amplifiers, and modulation/demodulation circuits.

In conclusion, satellite communication systems have revolutionized global connectivity, enabling seamless communication across vast distances. This subchapter provided an overview of these systems, including their fundamental principles, components, and applications. For electrical engineers, a solid understanding of satellite communication systems is crucial, as they play a key role in designing, developing, and optimizing these systems to ensure efficient and reliable communication.

Chapter 5: Telecommunication Technologies

Fiber Optic Communication

Fiber optic communication is a revolutionary technology that has transformed the field of telecommunications. In this subchapter, we will explore the principles, components, and applications of fiber optic communication systems.

Introduction to Fiber Optic Communication: Fiber optic communication is a method of transmitting information using light pulses through thin strands of glass or plastic known as optical fibers. These fibers act as a medium to transmit data over long distances at high speeds. The use of light instead of electrical signals allows for faster and more efficient data transmission.

Principles of Fiber Optic Communication: The foundation of fiber optic communication lies in the principle of total internal reflection. When light enters a fiber optic cable at a certain angle, it reflects off the inner walls of the fiber and continues to travel along its length. This phenomenon ensures that the light waves remain confined within the core of the fiber, minimizing signal loss and maintaining data integrity.

Components of Fiber Optic Communication: A typical fiber optic communication system consists of three main components: the transmitter, the fiber optic cable, and the receiver. The transmitter converts electrical signals into light pulses using a light source such as a laser or an LED. These light pulses are then transmitted through the fiber optic cable, which consists of a core surrounded by cladding material. Finally, the receiver converts the received light pulses back into electrical signals for further processing.

Applications of Fiber Optic Communication: Fiber optic communication has numerous applications in various fields, ranging from telecommunications to healthcare and data centers. It is widely used in long-distance communication networks, such as undersea cables and terrestrial fiber optic links, due to its ability to transmit data over vast distances without significant loss. Fiber optic cables are also used in high-speed internet connections, enabling faster and more reliable internet access. Additionally, fiber optic communication plays a crucial role in the healthcare industry, facilitating medical imaging techniques and telemedicine services.

Conclusion:
Fiber optic communication has revolutionized the field of telecommunications, offering faster, more reliable, and secure data transmission. Understanding the principles and components of fiber optic communication is essential for electrical engineers who work in the field of telecommunications. Mastering this technology opens up a world of possibilities for developing advanced communication systems and improving connectivity for everyone.

Principles of Fiber Optics

Fiber optics is a crucial technology in the field of telecommunications engineering. It involves the transmission of data, voice, and video signals using thin strands of glass or plastic, known as optical fibers. These fibers are capable of transmitting signals over long distances with minimal loss and interference, making them an ideal choice for high-speed and reliable communication networks.

Understanding the principles of fiber optics is essential for any electrical engineer working in the telecommunications industry. This subchapter aims to provide a comprehensive overview of the fundamental principles behind this technology.

Firstly, it is important to understand the concept of total internal reflection, which is the fundamental principle governing the transmission of light through optical fibers. Total internal reflection occurs when light traveling through a denser medium, such as glass or plastic, strikes the boundary with a less dense medium, such as air, at an angle greater than the critical angle. In this scenario, the light is reflected back into the denser medium, ensuring its propagation through the fiber.

Another key principle of fiber optics is the concept of light dispersion. Dispersion refers to the spreading out of different wavelengths of light as they travel through the fiber. This can lead to signal distortion and limit the transmission capacity of the fiber. Various techniques, such as the use of different types of fibers and dispersion compensation methods, are employed to mitigate the effects of dispersion.

Furthermore, the subchapter will delve into the different types of fiber optic cables and their characteristics. Single-mode fibers, which have a small core size, are designed to transmit a single mode of light, resulting in high bandwidth and low signal loss. On the other hand, multimode fibers have a larger core size and can transmit multiple modes of light simultaneously, making them suitable for shorter distance applications.

Additionally, the subchapter will cover various components of a fiber optic system, including transmitters, receivers, and connectors. It will explain the role of these components in the overall transmission process and discuss their design considerations.

Overall, this subchapter on the principles of fiber optics aims to equip electrical engineering professionals with the necessary knowledge to understand and work with this crucial technology. Whether you are involved in the design, installation, or maintenance of telecommunications networks, this subchapter will provide you with a solid foundation in fiber optics.

Fiber Optic Components and Systems

Introduction:
In today's digital age, where high-speed communication is essential, fiber optic technology has revolutionized the field of telecommunications. Fiber optic components and systems play a crucial role in transmitting vast amounts of data over long distances at lightning-fast speeds. This subchapter will delve into the fundamental concepts, components, and systems that make fiber optic technology possible.

Fundamental   Concepts:
To understand fiber optic components and systems, it is important to grasp the basic principles that underpin this technology. Fiber optics rely on the transmission of light signals through thin strands of glass or plastic fibers. These fibers are designed to carry data in the form of light pulses, which are sent and received using various components and systems.

Components:
Several key components are vital to the functioning of fiber optic systems. The first is the optical transmitter, which converts electrical signals into light signals. This process involves a light source, such as a laser or light-emitting diode (LED). The second component is the optical receiver, which receives and converts light signals back into electrical signals. It typically comprises a photodiode or a photodetector. A third crucial component is the optical fiber itself, responsible for guiding and transmitting the light signals with minimal loss.

Systems:

Fiber optic systems consist of more than just individual components. They involve complex arrangements of fibers, connectors, and other supporting elements. A fiber optic cable, for instance, comprises multiple fibers bundled together, enclosed in protective jackets. These cables can be installed underground, underwater, or overhead to facilitate long-distance communication.

Other essential systems include fiber optic connectors, which allow for easy connections between fibers, and couplers/splitters, which divide or combine optical signals. Wavelength-division multiplexing (WDM) systems enable the transmission of multiple signals simultaneously over the same fiber by using different wavelengths of light. Lastly, optical amplifiers boost the strength of signals during long-distance transmission, overcoming the inherent signal loss in fiber optic cables.

Conclusion:

Understanding fiber optic components and systems is crucial for electrical engineering professionals and enthusiasts alike. By harnessing the power of light signals, fiber optics has become the backbone of modern telecommunication systems. From optical transmitters and receivers to cables, connectors, and amplifiers, every component in a fiber optic system plays a vital role in delivering high-speed, reliable, and secure communication. The knowledge and mastery of these components and systems are essential for anyone involved in the field of electrical engineering and interested in the world of telecommunications.

Fiber Optic Network Design

In the rapidly evolving field of telecommunications, fiber optic networks have emerged as the backbone of modern communication systems. The design and implementation of these networks require a deep understanding of electrical engineering principles and practices. This subchapter aims to provide a comprehensive guide to mastering the art of fiber optic network design.

At its core, fiber optic network design involves the planning, layout, and optimization of the infrastructure required to transmit data over long distances using light pulses through optical fibers. These networks offer unparalleled advantages such as high bandwidth, low latency, and immunity to electromagnetic interference, making them the preferred choice for transmitting large amounts of data quickly and reliably.

To design an efficient fiber optic network, one must consider several key factors. Firstly, the geographic layout and topology play a crucial role in determining the network's architecture. Factors like distance, terrain, and existing infrastructure need to be carefully analyzed to ensure optimal performance and cost-effectiveness.

Next, understanding the various components that make up a fiber optic network is essential. These include optical transmitters, receivers, amplifiers, and multiplexers, among others. Each component must be selected based on the network's requirements and specifications, taking into account factors such as signal strength, dispersion, and noise.

Furthermore, the subchapter will delve into the different types of fiber optic cables available, such as single-mode and multi-mode fibers, and their associated characteristics. The selection of the appropriate cable type is critical to achieving the desired data transmission rates and distances.

Additionally, the subchapter will explore the crucial aspect of network scalability. As the demand for data continues to grow exponentially, designing a network that can accommodate future expansion is paramount. Techniques such as wavelength division multiplexing and fiber splicing will be covered to enable engineers to build scalable networks that can adapt to evolving needs.

Finally, the subchapter will address the importance of testing and maintenance in fiber optic network design. Regular inspections, cleaning, and troubleshooting are necessary to ensure the network's longevity and optimal performance.

In conclusion, the subchapter "Fiber Optic Network Design" aims to equip electrical engineering professionals and enthusiasts with a comprehensive understanding of the principles and practices involved in designing efficient fiber optic networks. By providing insights into various components, cable types, scalability, and maintenance, this guide will empower individuals to master the art of fiber optic network design and contribute to the advancement of telecommunications engineering.

Microwave Communication

Microwave communication is a fundamental aspect of modern electrical engineering and plays a crucial role in the field of telecommunications. This subchapter aims to provide a comprehensive understanding of microwave communication systems, their applications, and the underlying principles.

Microwave communication refers to the transmission of signals in the microwave frequency range, typically between 1 GHz and 300 GHz. These frequencies enable high-speed data transmission, making them ideal for various applications, including long-distance telecommunications, satellite communication, and wireless networks.

One of the key advantages of microwave communication is its ability to transmit large amounts of data over long distances with minimal signal degradation. Unlike traditional copper wire-based communication systems, microwave signals can travel through the air or space, allowing for efficient communication over vast areas. This makes microwave communication particularly suitable for connecting remote locations, such as rural areas or offshore platforms.

Microwave communication systems consist of various components, including transmitters, receivers, antennas, and waveguides. Transmitters convert electrical signals into microwave signals, which are then transmitted through antennas. Receivers located at the destination convert the received microwave signals back into electrical signals for further processing. The waveguides ensure the proper propagation of the microwave signals from the transmitter to the receiver.

Microwave communication finds extensive use in satellite communication systems. Satellites in geostationary or low Earth orbit act as relay stations, receiving signals from the ground and retransmitting them to other locations. This enables global communication, television broadcasting, and internet connectivity. Additionally, microwave communication is employed in microwave radio relay systems, which use a series of microwave towers to relay signals across long distances.

In the context of wireless networks, microwave communication is used for point-to-point and point-to-multipoint links. These links connect base stations, allowing for the distribution of data and voice signals to mobile devices. Microwave communication provides high capacity and low latency, making it ideal for supporting the ever-increasing demand for mobile data.

In conclusion, microwave communication is a vital aspect of electrical engineering, enabling efficient and high-speed data transmission over long distances. Its applications in telecommunications, satellite communication, and wireless networks make it a cornerstone of modern connectivity. Understanding the principles and components of microwave communication systems is essential for engineers working in the field of electrical engineering, ensuring they can design, optimize, and maintain these critical systems for the benefit of everyone.

Microwave Transmission Basics

Microwave transmission is a fundamental concept in the field of electrical engineering and plays a crucial role in modern telecommunications systems. This subchapter titled "Microwave Transmission Basics" from the book "Mastering Telecommunications Engineering: A Comprehensive Guide" aims to provide a comprehensive understanding of the principles behind microwave transmission, catering to a wide audience including students, professionals, and enthusiasts in electrical engineering.

Microwave transmission refers to the process of transmitting information, such as voice, data, or video, using high-frequency electromagnetic waves with wavelengths ranging from one meter to one millimeter. These waves are commonly referred to as microwaves. Microwave transmission has revolutionized the world of telecommunications due to its ability to transmit large amounts of information over long distances with minimal loss.

The subchapter begins by introducing the fundamental concepts of microwave transmission, including the properties of electromagnetic waves, such as frequency, wavelength, and amplitude. It explains how microwaves can be generated using specialized devices called microwave generators, which convert electrical energy into high-frequency electromagnetic waves.

The subchapter then delves into the various components of a microwave transmission system, such as antennas, waveguides, and amplifiers. It explains how antennas are used to transmit and receive microwave signals, and how they can be designed to achieve optimal performance. Waveguides, on the other hand, are used to guide the microwave signals from one point to another, minimizing losses along the transmission path.

Furthermore, the subchapter explores the concept of modulation, which is essential for encoding information onto microwave signals. It explains various modulation techniques, such as amplitude modulation (AM) and frequency modulation (FM), and their applications in microwave transmission.

Additionally, the subchapter provides an overview of the factors that affect microwave transmission, such as atmospheric conditions, interference, and line-of-sight requirements. It discusses how these factors can impact the quality and reliability of microwave communication links.

Overall, "Microwave Transmission Basics" is an essential subchapter in "Mastering Telecommunications Engineering: A Comprehensive Guide," providing a solid foundation in microwave transmission principles for individuals interested in the field of electrical engineering. Whether you are a student, a professional, or simply an enthusiast, this subchapter will equip you with the knowledge needed to understand and work with microwave transmission systems effectively.

Microwave Antenna Systems

Microwave antenna systems play a crucial role in modern telecommunications engineering, particularly in the field of electrical engineering. These systems are responsible for transmitting and receiving microwave signals over long distances, facilitating wireless communication and enabling the seamless transfer of data, voice, and video across the globe. In this subchapter, we will delve into the various components and technologies that make microwave antenna systems an indispensable part of our interconnected world.

Firstly, we will explore the fundamental principles of microwave transmission and propagation. Microwaves, as their name suggests, are electromagnetic waves with wavelengths ranging from one millimeter to one meter. They are highly directional and can be easily transmitted and received using specialized antennas. We will discuss the properties of microwaves and how they differ from other types of electromagnetic waves, such as radio waves and infrared waves.

Next, we will delve into the different types of microwave antennas commonly used in telecommunications engineering. These include parabolic antennas, horn antennas, and microstrip antennas, each with their own advantages and applications. We will explore the design, construction, and operation of these antennas, highlighting their unique features and capabilities.

Furthermore, we will discuss the different components that make up a microwave antenna system, such as feeders, waveguides, and reflectors. These components are crucial for ensuring efficient transmission and reception of microwave signals, minimizing losses and maximizing signal strength. We will delve into the design considerations and engineering principles behind these components, providing a comprehensive understanding of how they function within the overall system.

In addition, we will explore the latest advancements in microwave antenna technology, including phased array antennas and MIMO (Multiple-Input Multiple-Output) systems. These cutting-edge technologies have revolutionized the field of telecommunications engineering, allowing for faster data rates, increased capacity, and improved signal quality. We will discuss the principles behind these technologies and their potential applications in various fields, including wireless communication, satellite communication, and radar systems.

Finally, we will address the challenges and limitations of microwave antenna systems, such as signal interference, atmospheric attenuation, and environmental factors. We will discuss the strategies and techniques employed by engineers to overcome these challenges and ensure reliable and efficient communication.

Whether you are a student, a professional in the field of electrical engineering, or simply curious about the world of telecommunications, this subchapter on microwave antenna systems will provide you with a comprehensive understanding of the principles, technologies, and applications that underpin this vital aspect of modern communication.

Microwave Link Design

Microwave link design is a fundamental aspect of modern telecommunications engineering. It involves the planning, implementation, and optimization of microwave links to ensure reliable and efficient communication between various locations. This subchapter will provide a comprehensive overview of microwave link design principles, techniques, and considerations.

Microwave links play a crucial role in facilitating wireless communication over long distances. They utilize high-frequency electromagnetic waves to transmit voice, data, and multimedia information between two or more points. As a result, they are widely used in various applications, including mobile networks, satellite communications, and private data networks.

To design an effective microwave link, electrical engineers must consider several factors. Firstly, the geographical location of the link must be carefully assessed. Terrain features such as hills, mountains, and water bodies can significantly impact signal propagation. Engineers must conduct thorough path surveys and analyze the topography to determine the optimal location and alignment for the microwave link.

Additionally, engineers must consider the link budget, which involves assessing the power requirements, antenna gain, cable losses, and other factors that determine the overall signal strength and quality. By carefully analyzing these parameters, engineers can ensure that the microwave link operates within the desired performance limits.

Another critical aspect of microwave link design is frequency planning. Microwave frequencies are regulated by international standards, and engineers must select appropriate frequency bands that comply with these regulations. Interference from other nearby microwave links or wireless systems must also be considered during the frequency planning process.

Furthermore, engineers need to design suitable antennas for the microwave link. Antenna selection is crucial as it impacts the signal coverage, capacity, and link reliability. Engineers must consider factors such as antenna gain, beamwidth, polarization, and interference rejection capabilities.

Lastly, engineers must conduct thorough testing and optimization of the microwave link to ensure its performance meets the desired specifications. Field measurements, link quality tests, and signal-to-noise ratio assessments are some of the techniques used to evaluate and fine-tune the microwave link.

In conclusion, microwave link design is a complex yet crucial aspect of telecommunications engineering. By considering factors such as geography, link budget, frequency planning, antenna design, and optimization, electrical engineers can design microwave links that provide reliable and efficient communication over long distances. Understanding these principles is essential for anyone interested in the field of electrical engineering and the broader telecommunications industry.

Satellite Communication

Satellite communication has revolutionized the field of telecommunications, offering unparalleled opportunities for global connectivity. This subchapter explores the fundamental principles, technologies, and applications of satellite communication, providing a comprehensive guide to mastering this crucial aspect of electrical engineering.

Introduction to Satellite Communication Satellite communication refers to the use of artificial satellites in space to transmit and receive signals, enabling long-distance communication across vast areas of the Earth's surface. The advent of satellite communication has significantly expanded connectivity possibilities, allowing people from different corners of the world to communicate seamlessly.

Satellite Orbits and Frequencies Understanding the different types of satellite orbits is crucial in comprehending how satellite communication functions. This section explains the various orbit types, including geostationary orbit (GEO), medium Earth orbit (MEO), and low Earth orbit (LEO), and their respective advantages and limitations. Moreover, the allocation of frequency bands for satellite communication is discussed, highlighting the importance of frequency coordination and spectrum management.

Satellite Communication System Architecture This section delves into the architecture of a satellite communication system, encompassing the ground segment, space segment, and control segment. It explains the roles and functionalities of each segment, emphasizing the need for seamless coordination and synchronization for effective communication.

Satellite Link Design

Designing a satellite link involves several factors, including link budget calculations, antenna considerations, and modulation techniques. This subchapter provides a detailed overview of these crucial aspects, equipping electrical engineering professionals with the knowledge to optimize satellite link performance.

Applications of Satellite Communication

Satellite communication has found applications in various fields, ranging from telecommunications and broadcasting to weather forecasting and remote sensing. This section explores these applications, shedding light on the versatility and significance of satellite communication in our modern world.

Emerging Trends in Satellite Communication

The field of satellite communication is continuously evolving, with emerging trends presenting new opportunities and challenges. This subchapter explores cutting-edge advancements, such as high-throughput satellites, satellite constellations, and the integration of satellite communication with other technologies like 5G and Internet of Things (IoT).

Conclusion

Satellite communication is a vital component of modern electrical engineering, enabling global connectivity and a wide array of applications. This subchapter provides a comprehensive guide to mastering the principles, technologies, and emerging trends in satellite communication, equipping electrical engineering professionals with the knowledge and skills to excel in this dynamic field. Whether you are a student, researcher, or industry professional, understanding satellite communication is essential for navigating the ever-changing landscape of telecommunications.

Satellite Orbits and Coverage

Satellite Orbits and Coverage play a pivotal role in the world of telecommunications engineering. Understanding the different types of satellite orbits and their coverage areas is essential for electrical engineering professionals working in the field of telecommunications. In this subchapter, we will explore the fundamental concepts of satellite orbits and coverage, providing a comprehensive guide to mastering this crucial aspect of telecommunications engineering.

Satellites are positioned in various orbits around the Earth to ensure optimal coverage and communication capabilities. The most common types of satellite orbits include geostationary orbit (GEO), low Earth orbit (LEO), medium Earth orbit (MEO), and polar orbit. Each orbit has its own unique characteristics and advantages, making them suitable for specific applications.

Geostationary orbit, for instance, is positioned at an altitude of approximately 36,000 kilometers above the Earth's equator. Satellites in GEO orbits remain fixed relative to the Earth's surface, enabling continuous coverage over a specific region. This orbit is ideal for applications such as broadcasting, weather monitoring, and long-distance communications.

On the other hand, low Earth orbit (LEO) satellites are positioned at altitudes ranging from a few hundred to a few thousand kilometers. LEO satellites move at high speeds and complete an orbit around the Earth in a relatively short time. These satellites provide global coverage and are commonly used for applications like satellite phones, remote sensing, and Earth observation.

Medium Earth orbit (MEO) satellites, positioned at altitudes between GEO and LEO, offer a balance between coverage and signal latency. This orbit is commonly used for systems such as the Global Positioning System (GPS), where accurate positioning and timing are crucial.

Polar orbits, as the name suggests, pass over the Earth's poles. Satellites in polar orbits provide global coverage and are particularly useful for applications such as weather forecasting, environmental monitoring, and reconnaissance.

Understanding the coverage capabilities of different satellite orbits is vital for telecommunications engineers. Factors such as signal strength, latency, and coverage area need to be considered when designing and operating satellite-based systems.

In conclusion, the subchapter "Satellite Orbits and Coverage" provides an in-depth understanding of the various types of satellite orbits and their coverage characteristics. This knowledge is invaluable for electrical engineering professionals, allowing them to design and optimize satellite-based telecommunications systems for a wide range of applications. Whether you are a student, researcher, or industry professional, mastering the concepts of satellite orbits and coverage is essential for excelling in the field of telecommunications engineering.

Satellite Communication Systems

Satellite communication systems have revolutionized the way we connect and communicate across the globe. These systems play a critical role in various industries, including telecommunications, broadcasting, navigation, and remote sensing. This subchapter aims to provide a comprehensive understanding of satellite communication systems, their components, and their applications.

Introduction to Satellite Communication Systems Satellite communication systems involve the use of artificial satellites to transmit and receive signals over long distances. These systems rely on the principles of electromagnetic wave propagation and use satellites as relay stations to transmit data, voice, and video signals.

Components of Satellite Communication Systems Satellite communication systems consist of three main components: the ground segment, the space segment, and the control segment. The ground segment includes earth stations, antennas, and signal processing equipment. The space segment comprises satellites in geostationary or low-earth orbit, while the control segment involves ground-based control centers responsible for monitoring and controlling satellite operations.

Types of Satellites Satellites can be classified into different types based on their orbits, such as geostationary satellites and low-earth orbit satellites. Geostationary satellites orbit at an altitude of approximately 36,000 kilometers and remain stationary relative to the Earth's surface. Low-earth orbit satellites, on the other hand, orbit at much lower altitudes and offer advantages such as low latency and broader coverage.

Applications of Satellite Communication Systems Satellite communication systems find applications in various fields. In telecommunications, satellites enable long-distance communication, especially in remote areas where terrestrial infrastructure is limited. Broadcasting companies utilize satellites to transmit television and radio signals globally. Satellites also play a vital role in navigation systems like GPS, enabling accurate positioning and timing information. Additionally, satellite-based remote sensing provides valuable data for weather forecasting, disaster management, and environmental monitoring.

Challenges and Future Developments While satellite communication systems offer numerous benefits, they also face challenges like signal degradation due to atmospheric conditions and high initial deployment costs. However, advancements in technology, such as high-throughput satellites and the use of multiple frequency bands, are addressing these challenges and improving system performance.

In conclusion, satellite communication systems have revolutionized global connectivity and have become an integral part of various industries. This subchapter provides an overview of the components, types, applications, and challenges associated with satellite communication systems. As technology continues to advance, it is expected that satellite communication systems will play an even greater role in shaping the future of telecommunications and electrical engineering.

Earth Station Design and Operation

Introduction

In the ever-evolving field of telecommunications engineering, Earth station design and operation play a crucial role in ensuring seamless communication between different parts of the world. This subchapter aims to provide a comprehensive understanding of the principles, techniques, and best practices involved in designing and operating Earth stations. Whether you are an electrical engineering professional seeking to enhance your knowledge or an enthusiast interested in learning about modern telecommunication systems, this subchapter will serve as a valuable resource for you.

Overview   of   Earth   Stations
Earth stations act as key communication hubs that facilitate the transmission and reception of signals between satellites and ground-based systems. They enable various applications such as television broadcasting, internet connectivity, telephony, and data transfer. Understanding the key components of an Earth station, including antennas, transceivers, and control systems, is essential to comprehend their design and operation.

Design   Considerations
Effective Earth station design requires careful consideration of several factors. These include the selection of appropriate antenna types and sizes, frequency bands, receiver sensitivity, and power requirements. The subchapter will delve into these considerations, offering insights into the trade-offs involved and providing guidelines for optimal design decisions.

Earth   Station   Operations

Operating an Earth station involves managing a range of tasks to ensure reliable and efficient communication links. This subchapter will explore critical operational aspects such as signal acquisition, tracking, and pointing, as well as the monitoring and control of system parameters. It will also discuss common challenges faced during operations, including atmospheric conditions, interference, and system maintenance.

Advanced   Concepts

To cater to the needs of electrical engineering professionals and enthusiasts, this subchapter will also cover advanced topics related to Earth station design and operation. These may include emerging technologies such as software-defined radios, adaptive beamforming, and satellite constellations. By providing insights into cutting-edge developments, this subchapter aims to foster innovation and keep readers informed about the latest trends in the field.

Conclusion

Earth station design and operation are fundamental aspects of modern telecommunications engineering. Whether you are an electrical engineering professional seeking to expand your expertise or someone interested in gaining a comprehensive understanding of Earth station systems, this subchapter will serve as a valuable guide. By exploring the principles, techniques, and best practices in Earth station design and operation, readers will be equipped with the knowledge to contribute to the advancement of telecommunications systems and applications.

Chapter 6: Telecommunication Services and Applications

Voice Services

In the rapidly evolving field of telecommunications, voice services remain an integral part of our daily lives. From traditional landline telephones to the latest advancements in Voice over Internet Protocol (VoIP), the ability to communicate through voice has revolutionized the way we connect with one another. This subchapter will explore the fundamentals of voice services, their evolution, and their importance in the field of electrical engineering.

Voice services encompass a wide range of technologies and applications that enable the transmission of human speech over a network. Traditional voice services, such as the Public Switched Telephone Network (PSTN), have been the backbone of telecommunication systems for decades. However, with the advent of digital technologies, voice services have undergone a significant transformation.

One of the most notable advancements in voice services is the emergence of VoIP. This technology allows voice signals to be transmitted over the internet, providing cost-effective and flexible communication solutions. VoIP has revolutionized the telecommunications industry, allowing for features like video conferencing, call routing, and voicemail integration.

Electrical engineers play a vital role in the development and maintenance of voice services. They are responsible for designing, implementing, and optimizing the infrastructure that enables voice communication. From the design of network architectures to the development of compression algorithms, electrical engineers ensure the seamless transmission of voice signals.

In addition to the technical aspects, voice services also have a significant impact on society. They have bridged the gap between geographical boundaries, enabling people from different parts of the world to communicate effortlessly. Voice services have transformed industries such as customer service, healthcare, and education, allowing for remote interactions and access to information.

As the field of telecommunications continues to evolve, voice services will remain a fundamental component. Electrical engineers must stay updated with the latest advancements and technologies in voice services to effectively contribute to the industry. Understanding the principles behind voice transmission, the network protocols involved, and the integration of voice services with other telecommunications systems is crucial for electrical engineers.

In conclusion, voice services have revolutionized the way we communicate, creating a global network of connections. From traditional landline telephones to VoIP technologies, voice services have evolved to meet the changing needs of society. Electrical engineers play a crucial role in the development and maintenance of voice services, ensuring seamless communication across networks. By mastering the principles and technologies behind voice services, electrical engineers can continue to drive innovation in the field of telecommunications.

PSTN Services

PSTN, or Public Switched Telephone Network, refers to the traditional landline telephone system that has been in use for decades. In this subchapter, we will explore the various services offered by the PSTN and how they have shaped the field of telecommunications.

One of the fundamental services provided by the PSTN is voice communication. It allows individuals to make phone calls to anyone around the world using a network of interconnected switches and transmission lines. The PSTN ensures reliable and high-quality voice transmission, making it an integral part of modern society.

Apart from voice communication, the PSTN also offers supplementary services that enhance the overall user experience. These services include call waiting, call forwarding, caller ID, and voicemail. Call waiting allows users to receive incoming calls while already engaged in a conversation, while call forwarding enables calls to be redirected to another number. Caller ID displays the caller's information on the recipient's phone, and voicemail allows callers to leave messages when the recipient is unavailable. These features have become indispensable in our daily lives, enabling us to stay connected and manage our communications effectively.

Another essential service offered by the PSTN is fax transmission. Although email and other digital communication methods have gained popularity, fax machines are still widely used for transmitting documents securely. The PSTN supports fax services by converting the analog fax signals into digital signals for transmission over the network.

Moreover, the PSTN plays a crucial role in emergency services. When a user dials emergency services like 911, the call is routed through the PSTN to the respective emergency response center. This service ensures that prompt help can be provided during critical situations, saving lives and property.

While the PSTN has been the backbone of telecommunications for many years, its role is gradually evolving. With the advent of Voice over IP (VoIP) technology, which enables voice calls over the internet, traditional landline telephony is facing competition. However, the PSTN continues to serve as a reliable and secure communication infrastructure, especially in areas where internet connectivity may be limited or unreliable.

In conclusion, PSTN services have revolutionized the field of telecommunications, providing a reliable and efficient means of voice communication. The supplementary services offered by the PSTN enhance the user experience, while fax transmission and emergency services further demonstrate its importance in various sectors. As the telecommunications landscape continues to evolve, the PSTN remains a vital component of our interconnected world, ensuring seamless communication for everyone, particularly in the field of electrical engineering.

VoIP Services

In today's world, communication plays a vital role in every aspect of our lives. The advent of the internet has revolutionized the way we connect with each other, and one of the most significant advancements in this field is Voice over Internet Protocol (VoIP) services. This subchapter aims to provide a comprehensive understanding of VoIP services and their relevance in the field of electrical engineering.

VoIP services refer to the technology that allows voice communication to be transmitted over the internet rather than traditional telephone lines. This means that instead of using the traditional Public Switched Telephone Network (PSTN), VoIP utilizes the internet to transmit voice signals. This technology converts analog voice signals into digital data packets, which are then transmitted over the internet and reconverted into voice signals at the receiving end.

One of the key advantages of VoIP services is cost-effectiveness. Traditional telephone lines often incur high costs, especially for long-distance and international calls. VoIP eliminates these charges by utilizing the existing internet infrastructure, making it a much more affordable option for both businesses and individuals.

Additionally, VoIP services offer enhanced flexibility and scalability. With traditional telephone lines, adding or removing phone lines can be a cumbersome process. VoIP allows for easy scalability, enabling businesses to add or remove lines as per their requirements. Moreover, VoIP services provide a host of advanced features such as call forwarding, voicemail, and video conferencing, enhancing overall communication efficiency.

From an electrical engineering perspective, understanding VoIP services is essential for professionals in the field. Electrical engineers need to have a strong grasp of the underlying technology, including protocols like Session Initiation Protocol (SIP) and Real-Time Transport Protocol (RTP), which are vital for the successful implementation of VoIP services.

Furthermore, electrical engineers play a crucial role in the design and maintenance of the network infrastructure required for VoIP services. They are responsible for ensuring the quality of service, minimizing latency, and optimizing the network for seamless voice transmission.

In conclusion, VoIP services have revolutionized the way we communicate, offering cost-effectiveness, flexibility, and advanced features. For electrical engineers, understanding and mastering this technology is crucial to stay relevant in a rapidly evolving telecommunications landscape. This subchapter provides a comprehensive overview of VoIP services, addressing their significance and relevance in the field of electrical engineering, and equipping readers with the knowledge required to excel in this domain.

Unified Communications

Unified Communications (UC) is a concept that revolutionizes the way individuals and organizations communicate, collaborate, and share information. In the digital era, where seamless connectivity is paramount, UC provides a comprehensive set of tools and technologies to integrate various communication channels into a unified platform. This subchapter will explore the key aspects, benefits, and challenges of Unified Communications in the context of electrical engineering.

At its core, UC brings together various communication mediums such as voice, video, instant messaging, email, and data sharing into a single, cohesive system. This integration enables users to access and manage all their communication channels through a unified interface, regardless of the device or location they are using. By eliminating siloed communication systems, UC enhances productivity, efficiency, and collaboration among individuals and teams.

For electrical engineers, UC offers numerous advantages that streamline their work processes and enhance their ability to deliver innovative solutions. With UC, engineers can seamlessly communicate with colleagues, clients, and stakeholders in real-time, irrespective of their physical location. Engineers can collaborate on projects, share critical data and documents, and troubleshoot issues collectively, leading to faster decision-making and improved project outcomes.

Moreover, UC empowers electrical engineers with advanced features like presence, which allows them to see the availability and status of their colleagues, ensuring effective communication and reducing delays. Additionally, UC facilitates integration with other business applications and systems, enabling engineers to access relevant information, such as project specifications or equipment manuals, within their communication platform.

However, the implementation of UC in the electrical engineering field comes with its own set of challenges. Integration with legacy systems, ensuring data security, and managing the complexity of the UC infrastructure are some of the key hurdles that engineers need to address. Furthermore, as UC involves the convergence of multiple technologies, engineers must possess a comprehensive understanding of networking, protocols, and software applications to effectively design, deploy, and maintain UC systems.

In conclusion, Unified Communications revolutionizes the way electrical engineers communicate, collaborate, and share information. By integrating various communication channels into a unified platform, UC enhances productivity, efficiency, and collaboration within the electrical engineering field. While it offers numerous benefits, engineers must also overcome challenges related to integration, security, and infrastructure management. Mastering UC is essential for electrical engineers to stay ahead in the fast-paced and interconnected world of modern telecommunications.

Data Services

In today's digital age, data has become the lifeblood of our interconnected world. From sending a simple email to streaming high-definition videos, data services play a crucial role in our daily lives. This subchapter will provide a comprehensive overview of data services in the field of telecommunications engineering, exploring the various aspects and technologies involved.

Data services encompass a wide range of applications, including internet access, cloud computing, and data storage. These services enable individuals and businesses to communicate, collaborate, and access information seamlessly. With the increasing demand for high-speed and reliable data connectivity, telecommunications engineers play a vital role in designing, implementing, and maintaining data networks.

One of the fundamental components of data services is internet access. Telecommunications engineers are responsible for designing and deploying the infrastructure required to connect users to the internet. This includes the installation of fiber optic cables, routers, switches, and other networking equipment. The subchapter will delve into the different types of internet connections, such as broadband, DSL, cable, and wireless, providing a comprehensive understanding of their strengths and limitations.

Cloud computing has revolutionized the way data is stored and processed. Telecommunications engineers are at the forefront of developing and managing cloud-based data services. The subchapter will explore the concept of cloud computing, its benefits, and the underlying technologies, such as virtualization and distributed computing. Additionally, it will discuss the challenges and considerations involved in ensuring data security and privacy in cloud environments.

Data storage is another critical aspect of data services. With the exponential growth of digital information, efficient and reliable storage solutions are paramount. Telecommunications engineers are experts in designing and implementing storage systems, including technologies like solid-state drives (SSDs), network-attached storage (NAS), and storage area networks (SANs). The subchapter will delve into these technologies, their advantages, and their applications in various contexts.

In conclusion, data services are vital for our interconnected world, enabling seamless communication, collaboration, and access to information. Telecommunications engineers play a crucial role in designing, implementing, and maintaining the infrastructure required for these services. This subchapter aims to provide a comprehensive understanding of data services, covering topics such as internet access, cloud computing, and data storage. Whether you are an electrical engineering student, a telecommunications professional, or simply interested in the field, this subchapter will equip you with the knowledge needed to navigate the world of data services.

Internet Services

In today's digital age, the internet has become an integral part of our lives. It has revolutionized the way we communicate, work, and access information. Internet services have become a necessity for individuals, businesses, and even governments. This subchapter aims to provide a comprehensive overview of internet services, their significance, and their relevance in the field of electrical engineering.

Internet services encompass a wide range of applications and technologies that enable users to connect, communicate, and share information over the internet. These services include web browsing, email, social media, online gaming, video streaming, cloud computing, and much more. Each service is designed to meet specific needs and requirements of users, whether it's staying connected with friends and family, conducting business operations, or accessing entertainment content.

In the field of electrical engineering, internet services play a crucial role in the development and deployment of various technologies and systems. From the design and implementation of network infrastructures to the development of communication protocols and security mechanisms, understanding internet services is essential for electrical engineers.

One of the key aspects of internet services is their reliance on networks. Electrical engineers need to have a deep understanding of network architectures, protocols, and technologies to ensure the efficient and secure transmission of data over the internet. This includes knowledge of TCP/IP, Ethernet, wireless networks, routers, switches, and firewalls.

Internet services also require robust and reliable infrastructure to support their operations. Electrical engineers are responsible for designing and maintaining data centers, which house the servers and equipment necessary to provide internet services. They must consider factors such as power consumption, cooling systems, and redundancy to ensure uninterrupted service availability.

Furthermore, electrical engineers play a vital role in developing innovative internet services. They are involved in the design and implementation of new communication technologies, such as 5G networks, Internet of Things (IoT), and virtual reality (VR) applications. These advancements not only enhance the user experience but also drive economic growth and societal development.

In conclusion, internet services have become an integral part of our lives, transforming the way we communicate, work, and access information. For electrical engineers, understanding internet services is crucial to design, develop, and maintain the necessary infrastructure and technologies that enable seamless connectivity and efficient data transmission. By mastering internet services, electrical engineers can contribute to the continuous evolution and improvement of the internet, benefiting individuals, businesses, and society as a whole.

Cloud Computing

Cloud computing is a revolutionary technology that has transformed the field of electrical engineering. It has become an integral part of our everyday lives, impacting the way we store, access, and process data. In this subchapter, we will explore the concept of cloud computing, its benefits, and its applications in the field of electrical engineering.

Cloud computing refers to the delivery of computing services, including storage, databases, software, and networking over the internet. Instead of relying on local servers or personal computers, cloud computing allows users to access and utilize these resources remotely. This technology is based on a network of remote servers hosted on the internet, commonly known as the cloud.

One of the key advantages of cloud computing is its scalability. It allows electrical engineers to easily scale up or down their computing resources based on their needs. This flexibility eliminates the need for expensive hardware investments and provides a cost-effective solution for companies and individuals. Additionally, cloud computing offers high availability, ensuring that electrical engineers can access their data and applications anytime, anywhere.

Electrical engineers can leverage cloud computing in various ways. For instance, it enables them to store and process large amounts of data efficiently. Cloud-based storage services provide a secure and reliable solution for backing up critical data, reducing the risk of data loss. Furthermore, electrical engineers can use cloud platforms to develop and deploy applications, eliminating the need for infrastructure management. This allows engineers to focus on their core tasks and accelerates the development process.

The use of cloud computing in electrical engineering also promotes collaboration and remote work. With cloud-based collaboration tools, engineers can work together on projects in real-time, regardless of their physical location. This enhances productivity and fosters innovation within the field.

However, it is vital to consider the security implications of cloud computing. Electrical engineers must ensure that their data is protected from unauthorized access and cyber threats. Implementing robust security measures, such as encryption and access controls, is crucial to safeguard sensitive information.

In conclusion, cloud computing has revolutionized the field of electrical engineering by providing scalable, cost-effective, and flexible computing resources. It enables engineers to store, process, and access data efficiently, promotes collaboration, and streamlines development processes. As this technology continues to evolve, electrical engineers must stay updated with the latest advancements and security measures to fully leverage the benefits of cloud computing.

Virtual Private Networks (VPNs)

In today's interconnected world, where communication and data exchange are at the heart of every industry, the need for secure and reliable network connections has become paramount. This is where Virtual Private Networks (VPNs) come into play. VPNs have revolutionized the field of telecommunications engineering, providing a secure and private means of transmitting data over public networks.

For EVERY ONE interested in Electrical Engineering, understanding VPNs is crucial. A VPN essentially creates a private network connection over a public network infrastructure, such as the internet. By employing various encryption and authentication methods, VPNs ensure that data transmitted between two endpoints remains confidential and protected from unauthorized access.

VPNs offer numerous benefits, making them a vital tool for businesses and individuals alike. Firstly, they enable remote access to private networks, allowing employees to securely connect to their work networks from anywhere in the world. This flexibility has become even more critical with the rise of remote work and the need for seamless collaboration across geographically dispersed teams.

Another advantage of VPNs is their ability to bypass geographical restrictions. By connecting to a VPN server located in a different country, users can access content that may be restricted or censored in their own region. This feature has made VPNs popular among individuals who seek unrestricted access to the internet and value their online privacy.

Furthermore, VPNs can enhance the security of public Wi-Fi networks. When connected to a public hotspot, such as those found in cafes or airports, data transmitted without encryption can be intercepted by malicious actors. By using a VPN, all data passing through the network is encrypted, ensuring that sensitive information remains protected.

It is important to note that while VPNs provide an additional layer of security, they are not foolproof. Users must select reputable VPN providers and avoid engaging in potentially risky online behavior. Additionally, VPNs may introduce latency and impact network performance, especially when dealing with large amounts of data.

In conclusion, Virtual Private Networks (VPNs) have become indispensable tools for ensuring secure and private communication over public networks. For EVERY ONE interested in Electrical Engineering, understanding the concepts and applications of VPNs is crucial. From enabling remote access to private networks to bypassing geographical restrictions and enhancing public Wi-Fi security, VPNs offer a wide range of advantages. However, users must exercise caution and select reliable providers to ensure the effectiveness of VPNs in protecting sensitive data.

Mobile Services

In today's fast-paced world, mobile services have become an integral part of our daily lives. From making phone calls and sending text messages to accessing the internet and using a wide range of applications, mobile services have revolutionized the way we communicate and interact with the world around us.

Mobile services, also known as wireless services, encompass a broad range of technologies and applications that enable wireless communication between devices. These services are made possible through the use of cellular networks, which consist of a network of base stations that transmit and receive signals to and from mobile devices.

One of the key advantages of mobile services is their portability. Unlike traditional landline telephones, mobile devices can be carried anywhere, allowing users to stay connected on the go. This mobility has transformed the way we conduct business, stay in touch with loved ones, and access information.

Mobile services offer a plethora of features and applications that cater to various needs and preferences. Voice calls remain a fundamental aspect of mobile services, facilitating real-time communication between individuals. Text messaging, on the other hand, allows for quick and convenient written communication, making it an essential tool for everyday conversations.

In addition to voice calls and text messages, mobile services also provide access to the internet. With the advent of smartphones, users can browse websites, check emails, and engage in social media platforms, all from the palm of their hand. This constant connectivity has opened up a world of opportunities, enabling us to stay updated with news, access online services, and connect with people from all corners of the globe.

Moreover, mobile services have given rise to a vast array of applications that cater to different niches and interests. From productivity tools and entertainment apps to health and fitness trackers, the possibilities are endless. These applications not only enhance our daily lives but also contribute to the growing field of electrical engineering, as they constantly push the boundaries of innovation and technology.

In conclusion, mobile services have revolutionized the way we communicate and access information. Their portability, versatility, and wide range of applications have made them an indispensable part of our daily lives. Whether you are an electrical engineering enthusiast or simply curious about the world of telecommunications, understanding mobile services is crucial in navigating the ever-evolving landscape of wireless communication.

Mobile Network Architectures

Mobile network architectures play a crucial role in the field of telecommunications engineering. With the rapid advancement of mobile technology, it is essential to understand the different architectures that underpin mobile networks. This subchapter aims to provide a comprehensive overview of mobile network architectures, their components, and their significance.

Mobile network architectures refer to the structures and frameworks that enable the functioning of mobile communication systems. These architectures are designed to facilitate the seamless transmission of voice, data, and multimedia content over wireless networks. They are built upon a combination of hardware and software components, each serving a specific purpose in the network.

One of the most widely used mobile network architectures is the GSM (Global System for Mobile Communications) architecture. GSM is a standard for digital cellular communication used by mobile phones and other mobile devices. It encompasses various components such as base stations, mobile switching centers, and the Home Location Register (HLR). Understanding the GSM architecture is fundamental to comprehending the foundations of mobile communication.

Another prominent mobile network architecture is the CDMA (Code Division Multiple Access) architecture. CDMA uses a spread spectrum technique to enable multiple users to share the same frequency band simultaneously. It employs base stations, mobile switching centers, and a Mobile Station (MS) to facilitate communication. CDMA is widely used in 3G and 4G networks and has significantly contributed to the evolution of mobile technology.

In recent years, the emergence of 5G technology has revolutionized mobile network architectures. 5G networks operate on a completely new architecture known as the Service-Based Architecture (SBA). This architecture is designed to provide ultra-low latency, high bandwidth, and support for massive IoT (Internet of Things) connectivity. It introduces concepts such as Network Function Virtualization (NFV) and Software-Defined Networking (SDN), which enhance the flexibility and scalability of mobile networks.

Understanding mobile network architectures is crucial for electrical engineering professionals working in the telecommunications industry. It enables them to design, deploy, and maintain efficient mobile networks. Furthermore, a solid understanding of these architectures is essential for troubleshooting network issues and optimizing network performance.

In conclusion, mobile network architectures are the backbone of mobile communication systems. They provide the necessary infrastructure and components for seamless wireless communication. This subchapter has provided an overview of prominent architectures like GSM, CDMA, and 5G, emphasizing their significance in the field of electrical engineering. By mastering these architectures, professionals can contribute to the advancement of mobile technology and enhance the user experience for everyone.

Mobile Applications

Mobile applications, also known as mobile apps, have become an integral part of our daily lives. From social media platforms to digital banking services, these apps have revolutionized the way we communicate, work, and interact with the world around us. In this subchapter, we will explore the world of mobile applications and their significance in the field of electrical engineering.

Mobile applications are software programs designed to run on mobile devices such as smartphones and tablets. They are developed to perform specific tasks, provide information, or entertain users. With the rapid advancement of technology, the capabilities of mobile apps have expanded exponentially, making them indispensable tools for both personal and professional use.

In the field of electrical engineering, mobile applications play a crucial role in various aspects. They enable engineers to access important data and resources on the go, making their work more efficient and convenient. For example, there are numerous mobile apps that provide real-time monitoring and control of different electrical systems, allowing engineers to remotely manage and troubleshoot issues. These apps not only save time but also improve the overall productivity of engineers.

Moreover, mobile applications have opened up new avenues for innovation and research in electrical engineering. Developers can create apps that simulate electrical circuits, analyze data, and even assist in the design process. These apps provide a platform for engineers to experiment and test their ideas in a virtual environment before implementing them in the real world.

Furthermore, mobile applications have also contributed to the advancement of smart homes and Internet of Things (IoT) devices. Through intuitive interfaces and connectivity features, these apps allow users to control and monitor various electrical appliances and devices remotely. This integration of mobile apps with electrical engineering has paved the way for a more connected and automated future.

In conclusion, mobile applications have brought about significant changes in the field of electrical engineering. They have enhanced the productivity of engineers, facilitated innovation, and enabled the creation of smart and interconnected systems. As technology continues to evolve, the role of mobile apps in electrical engineering will only continue to grow, offering new opportunities and challenges for professionals in the field.

Mobile Security

In today's digital age, mobile devices have become an indispensable part of our lives. From smartphones to tablets, these devices have revolutionized the way we communicate, work, and access information. However, with this increased connectivity and convenience comes the need for robust mobile security measures.

Mobile security refers to the protective measures implemented to safeguard personal and sensitive information stored on mobile devices from unauthorized access, malware, data breaches, and other cyber threats. As the reliance on mobile devices continues to grow, so does the importance of understanding and implementing effective mobile security practices.

This subchapter aims to provide a comprehensive overview of mobile security, covering various aspects relevant to electrical engineering professionals and anyone interested in understanding the principles behind securing mobile devices.

The subchapter begins by exploring the unique challenges posed by mobile devices in terms of security. Mobile devices are highly vulnerable due to their portability, wireless connectivity, and the increasing sophistication of cyber threats. The chapter delves into the potential risks associated with mobile devices, such as data leakage, unauthorized access, and mobile malware.

Next, the subchapter discusses the fundamental principles of mobile security. It covers topics such as encryption, authentication, secure coding practices, and the importance of regular updates and patches. Additionally, it explores the role of mobile device management

(MDM) solutions in enforcing security policies and ensuring data protection.

Furthermore, the subchapter provides an overview of the different types of mobile security threats and attack vectors. It delves into common attack techniques such as phishing, malware, and network-based attacks. It also highlights the importance of user awareness and education in preventing social engineering attacks and maintaining good mobile security hygiene.

The subchapter concludes with a discussion on emerging trends and future challenges in mobile security. It highlights the need to adapt security measures to keep up with evolving technologies such as 5G and the Internet of Things (IoT). Additionally, it emphasizes the importance of collaboration between electrical engineering professionals, software developers, and cybersecurity experts to develop innovative solutions that can effectively combat mobile security threats.

In summary, this subchapter on mobile security provides a comprehensive guide for electrical engineering professionals and individuals interested in understanding the principles and best practices for securing mobile devices. By implementing the knowledge gained from this subchapter, readers can enhance their mobile security posture and protect their personal and sensitive information in an increasingly connected world.

Chapter 7: Telecommunication Engineering and Design

Telecommunication System Design Principles

In the realm of electrical engineering, telecommunication systems play a crucial role in connecting people and enabling the exchange of information across vast distances. The design of these systems requires meticulous planning and adherence to a set of principles that ensure reliable and efficient communication. This subchapter delves into the fundamental design principles that underpin the development of telecommunication systems, providing a comprehensive guide for both seasoned professionals and those entering the field.

One of the foremost principles in telecommunication system design is scalability. As technology continues to evolve at a rapid pace, it is essential to design systems that can accommodate future growth and advancements. By incorporating scalable components and infrastructure, telecommunication systems can adapt to increasing demands without requiring significant reconfiguration or investment.

Another critical principle is redundancy. Given the importance of uninterrupted communication, designing systems with redundant components and backup solutions is paramount. Redundancy ensures that in the event of a failure or disruption, there are alternative routes and resources available to maintain connectivity, minimizing downtime and enhancing system reliability.

Security is an equally vital principle in telecommunication system design. With the increasing prevalence of cyber threats, protecting sensitive information and preventing unauthorized access is of utmost importance. Designing systems with robust security measures, such as encryption protocols and firewalls, helps safeguard data and maintain the privacy of users.

Efficiency is another principle that guides telecommunication system design. By optimizing network infrastructure, reducing signal loss, and utilizing bandwidth effectively, efficient systems can transmit data at higher speeds while minimizing energy consumption. This not only enhances user experience but also contributes to a sustainable and environmentally friendly approach to telecommunications.

Interoperability is also a key principle that ensures seamless communication across different systems and devices. Designing systems with standardized interfaces and protocols allows for compatibility and ease of integration, enabling different components and technologies to work together harmoniously.

Finally, usability and user experience are vital considerations. Designing telecommunication systems that are intuitive, user-friendly, and accessible to a wide range of individuals ensures that technology is inclusive and usable for all.

By understanding and applying these telecommunication system design principles, electrical engineers can create robust, scalable, secure, and efficient systems that empower global connectivity and shape the future of communication. Whether you are a seasoned professional or new to the field, this subchapter provides invaluable insights and guidance for mastering telecommunication engineering.

Network Planning and Optimization

In the rapidly evolving field of telecommunications engineering, network planning and optimization play a crucial role in ensuring the smooth and efficient functioning of communication systems. This subchapter aims to provide a comprehensive guide to the principles, methodologies, and best practices involved in network planning and optimization, addressing a diverse audience ranging from students and professionals in electrical engineering to enthusiasts interested in the field.

Network planning involves the strategic design and layout of telecommunications networks to meet specific requirements. It encompasses various aspects, including capacity planning, coverage planning, and topology design. By carefully considering factors such as traffic patterns, user demands, and future growth projections, network planners can ensure that the network infrastructure is designed to handle current and future needs efficiently.

Optimization, on the other hand, focuses on enhancing the performance of existing networks. It involves continuous monitoring, analysis, and fine-tuning of network parameters to maximize efficiency, reliability, and quality of service. Optimization techniques encompass a wide range of activities, such as antenna optimization, frequency planning, interference management, and resource allocation.

This subchapter will delve into the fundamental concepts and methodologies used in network planning and optimization. It will cover topics such as network modeling, traffic analysis, propagation modeling, link budget calculations, and network performance evaluation. Additionally, it will explore advanced optimization techniques, including machine learning algorithms and artificial intelligence-based approaches.

To cater to the diverse audience of electrical engineering students, professionals, and enthusiasts, this subchapter will provide a balanced blend of theoretical knowledge and practical insights. Real-world examples, case studies, and industry best practices will be incorporated to facilitate a deeper understanding of the subject matter.

By mastering the principles and techniques of network planning and optimization, readers will be equipped with the necessary skills to design, deploy, and maintain robust telecommunications networks. The knowledge gained from this subchapter will empower them to tackle complex challenges in the field and contribute to the advancement of the telecommunications industry.

Whether you are a student embarking on a career in electrical engineering or a professional seeking to enhance your knowledge, this subchapter will serve as an invaluable resource in mastering the intricacies of network planning and optimization in the dynamic world of telecommunications engineering.

Transmission Line Design

In the vast field of electrical engineering, transmission line design holds immense importance as it forms the backbone of modern telecommunication systems. This subchapter aims to provide a comprehensive understanding of transmission line design, covering its principles, components, and applications.

Transmission lines serve as crucial conduits for the transmission of electrical signals over long distances, ensuring efficient and reliable communication networks. Understanding the design principles of transmission lines is essential for electrical engineers involved in the planning, installation, and maintenance of telecommunication systems.

This subchapter begins by introducing the fundamental concepts of transmission line design, including the various types of transmission lines, such as coaxial cables, waveguides, and optical fibers. It explores the unique characteristics and advantages of each type, allowing engineers to make informed decisions based on specific application requirements.

The subsequent sections delve into the components that constitute a transmission line, discussing topics such as conductors, insulators, dielectrics, and shielding. It highlights the importance of selecting appropriate materials and designs to ensure optimal signal transmission, minimal losses, and protection against external interference.

Furthermore, this subchapter provides insights into the design considerations for transmission line impedance matching, a critical aspect for minimizing signal reflections and maximizing power transfer efficiency. It covers topics such as characteristic impedance, standing waves, and the Smith chart, equipping engineers with the knowledge to optimize transmission line performance.

Additionally, the content explores the impact of transmission line design on signal attenuation, distortion, and dispersion, as well as the techniques employed to mitigate these effects. It discusses equalization, amplification, and error correction methods, enabling engineers to maintain signal integrity over long-distance transmission lines.

Lastly, this subchapter sheds light on the practical applications of transmission line design, including telecommunications networks, data centers, and high-speed digital communication systems. It emphasizes the importance of considering factors such as bandwidth, data rate, and signal quality when designing transmission lines for specific applications.

In conclusion, the subchapter on transmission line design serves as an indispensable resource for electrical engineers, providing a comprehensive guide to understanding the principles, components, and applications of transmission lines. By mastering the principles outlined in this subchapter, engineers can design efficient and reliable telecommunication systems that meet the demands of the ever-evolving field of electrical engineering. Whether one is a student, professional engineer, or an enthusiast, this subchapter offers valuable insights into the world of transmission line design.

Antenna System Design

In the ever-evolving world of telecommunications, the design and implementation of an efficient and effective antenna system is crucial. Whether you are a seasoned electrical engineer or someone with a general interest in the field, understanding the key principles and considerations behind antenna system design is essential.

This subchapter of "Mastering Telecommunications Engineering: A Comprehensive Guide" aims to demystify the complex world of antenna systems and provide readers with a comprehensive understanding of the design process. From basic concepts to advanced techniques, we will cover it all.

To begin, we will delve into the fundamental principles of antenna design, including the various types of antennas and their applications. We will explore the differences between omnidirectional and directional antennas, discussing their advantages and disadvantages. Additionally, we will examine specialized antennas such as Yagi-Uda, helical, and phased array antennas.

Next, we will explore the factors that influence antenna performance. These factors include antenna gain, radiation patterns, impedance matching, polarization, and bandwidth. By understanding these concepts and their interplay, engineers can optimize an antenna system to achieve maximum performance.

Furthermore, we will discuss the importance of proper installation and placement of antennas. Factors such as height, location, and obstructions can significantly impact signal reception and transmission. We will provide guidelines and best practices for positioning antennas to minimize interference and maximize signal strength.

Additionally, we will cover antenna system integration, addressing topics such as feeder cables, connectors, and lightning protection. Understanding how these components interact within the system is vital to ensure optimal performance and longevity.

Lastly, we will explore emerging trends and technologies in antenna system design. With the advent of 5G, the Internet of Things (IoT), and satellite communications, engineers must stay up to date with the latest advancements. We will discuss topics such as MIMO (Multiple-Input Multiple-Output) antennas and beamforming techniques, which are revolutionizing the field.

By the end of this subchapter, readers will have a solid understanding of antenna system design principles and techniques. Whether you are an electrical engineer looking to deepen your knowledge or someone curious about the world of telecommunications, this chapter will provide you with the tools to master antenna system design and enhance your understanding of this crucial aspect of modern technology.

Telecommunication Infrastructure Design

In the rapidly evolving world of telecommunications, the design of a robust and efficient telecommunication infrastructure plays a crucial role in enabling the seamless transfer of data, voice, and multimedia. This subchapter explores the fundamental principles and techniques involved in the design of a telecommunication infrastructure, catering to a wide audience including electrical engineering professionals, students, and enthusiasts.

The subchapter begins with an introduction to the importance of telecommunication infrastructure design in today's interconnected world. It emphasizes the need for reliable and scalable networks that can handle the ever-increasing demand for data transmission, while ensuring minimal latency and maximum security.

The subchapter then delves into the key components of telecommunication infrastructure design, starting with the physical layer. It covers topics such as site selection, tower design, and cabling methodologies, providing a comprehensive overview of the various considerations that go into creating a robust physical infrastructure.

Moving on, the subchapter explores the design of the network architecture, focusing on the different types of networks, including LAN, WAN, and wireless networks. It examines the principles of network topology, protocol selection, and network equipment placement, offering insights into creating efficient and scalable networks.

Furthermore, the subchapter discusses the design of the telecommunication system, covering aspects such as signal processing, modulation techniques, and multiplexing methods. It highlights the importance of optimizing signal quality and bandwidth utilization, as well as the integration of emerging technologies such as fiber optics and software-defined networking.

To provide a holistic understanding, the subchapter also addresses the design considerations for telecommunication services, including voice, data, and multimedia transmission. It explores topics such as Quality of Service (QoS), traffic engineering, and service-level agreements, enabling the audience to design networks that meet the diverse needs of end-users.

Throughout the subchapter, practical examples, case studies, and illustrations are provided to enhance the learning experience. Additionally, discussions on emerging trends and technologies, such as 5G and Internet of Things (IoT), provide the audience with valuable insights into the future of telecommunication infrastructure design.

Whether you are an electrical engineering professional seeking to enhance your knowledge or a student aiming to delve into the world of telecommunication, this subchapter serves as a comprehensive guide, equipping you with the necessary tools and techniques to design efficient and robust telecommunication infrastructures.

Chapter 8: Emerging Trends in Telecommunications Engineering

Internet of Things (IoT)

The Internet of Things (IoT) is a rapidly growing field in the realm of Electrical Engineering that has revolutionized the way we interact with the world around us. In this subchapter, we will explore the concept of IoT and its implications for both professionals and everyday individuals.

IoT refers to the network of physical devices, vehicles, appliances, and other objects embedded with sensors, software, and connectivity, enabling them to collect and exchange data. This interconnectedness allows for seamless communication between devices, creating a web of intelligent systems that can be remotely monitored, controlled, and automated.

The impact of IoT is far-reaching and touches various aspects of our lives. From smart homes equipped with automated lighting and heating systems, to wearable devices that track our health and fitness, IoT is transforming the way we live and interact with technology. Moreover, IoT has found applications in industries such as agriculture, healthcare, transportation, and manufacturing, where it has streamlined processes, increased efficiency, and enhanced productivity.

For electrical engineers, understanding the principles and technologies behind IoT is essential. It involves a multidisciplinary approach, combining knowledge of electronics, wireless communication, sensor networks, and data analytics. This subchapter will delve into the various components of IoT, including sensors, actuators, communication protocols, and cloud computing, providing a comprehensive guide to mastering the field of telecommunications engineering.

Furthermore, we will explore the challenges and opportunities that arise with the proliferation of IoT. Security and privacy concerns are major considerations, as the interconnected nature of IoT opens up vulnerabilities to cyber attacks. Engineers must develop robust security measures to protect sensitive data and ensure the integrity of connected systems.

In conclusion, the Internet of Things is a transformative force in the world of Electrical Engineering, impacting not only professionals in the field but also everyday individuals. This subchapter will equip readers with the necessary knowledge and skills to navigate the complexities of IoT, enabling them to harness its potential and contribute to the advancement of this rapidly evolving field. Whether you are a seasoned engineer or a curious individual with a passion for technology, this subchapter will serve as a valuable resource in mastering the realm of telecommunications engineering.

5G Communication Systems

In recent years, there has been a significant buzz surrounding the development and deployment of 5G communication systems. This revolutionary technology is set to transform the way we connect and communicate, opening up new possibilities for industries and individuals alike. In this subchapter, we will delve into the world of 5G, exploring its key features, benefits, and the impact it is likely to have on the field of electrical engineering.

5G, or the fifth generation of wireless technology, promises faster speeds, lower latency, and greater capacity than its predecessors. It is designed to handle the increasing demands of our digitally connected world, where everything from smart homes to autonomous vehicles requires a seamless and reliable network connection. With 5G, users can expect download speeds of up to 10 gigabits per second, enabling near-instantaneous data transfer and a more immersive user experience.

One of the most significant advancements brought about by 5G is the ability to support the Internet of Things (IoT) on a large scale. This means that billions of devices, ranging from sensors and wearables to industrial machinery, can be connected and communicate with each other. This opens up a world of possibilities for electrical engineers, who can design and implement innovative solutions that leverage the power of the IoT to improve efficiency, optimize processes, and create new business models.

Moreover, 5G will enable the widespread adoption of emerging technologies such as virtual reality (VR), augmented reality (AR), and autonomous systems. These technologies heavily rely on low latency and high bandwidth, which 5G can provide. Electrical engineers will play a crucial role in developing the infrastructure and devices needed to support these applications, paving the way for a more interconnected and immersive future.

However, the implementation of 5G communication systems also presents its fair share of challenges. The deployment of a dense network of small cells, the integration of multiple frequency bands, and the management of interference are just a few of the hurdles that electrical engineers will need to overcome. These challenges require a deep understanding of wireless communication principles, signal processing techniques, and network optimization strategies.

In conclusion, 5G communication systems are set to revolutionize the field of electrical engineering by providing faster speeds, lower latency, and greater capacity. This technology opens up new possibilities for the IoT, virtual reality, augmented reality, and autonomous systems. However, it also presents challenges that electrical engineers must face head-on. By mastering the intricacies of 5G, electrical engineers will be at the forefront of innovation, shaping the future of telecommunications and beyond.

Artificial Intelligence in Telecommunications

Artificial Intelligence (AI) has emerged as a transformative technology in various industries, and the telecommunications sector is no exception. In recent years, AI has revolutionized the way telecommunications services are delivered, enhancing efficiency, improving customer experience, and enabling new capabilities. This subchapter explores the applications of AI in telecommunications, highlighting its impact on the field of electrical engineering.

AI-powered virtual assistants have become a common feature in telecommunication services, providing customers with personalized support and improving overall customer satisfaction. These virtual assistants leverage natural language processing and machine learning algorithms to understand and respond to customer queries, helping to troubleshoot technical issues, recommend products and services, and even perform basic account management tasks.

Beyond customer service, AI is also being utilized to optimize network performance and maintenance. Telecommunication networks generate massive amounts of data, and AI algorithms can analyze this data to identify patterns, predict failures, and proactively resolve issues. By using AI techniques such as anomaly detection and predictive maintenance, electrical engineers can ensure that the network remains operational and downtime is minimized.

Furthermore, AI is enabling the development of advanced network routing and optimization algorithms. These algorithms can dynamically allocate network resources based on real-time demands, ensuring efficient utilization of available capacity and reducing congestion. This capability is particularly crucial in the era of 5G networks, where the vast increase in connected devices and data traffic necessitates intelligent network management.

In addition, AI is playing a pivotal role in the security of telecommunications systems. With the proliferation of cyber threats, electrical engineers are constantly seeking ways to enhance network security. AI-powered security systems can automatically detect and respond to potential threats, analyze network traffic for malicious activity, and even predict and prevent future attacks. By leveraging AI, electrical engineers can strengthen the resilience and integrity of telecommunication networks.

As AI continues to evolve, its potential in the telecommunications industry remains vast. From improving customer service to optimizing network performance and enhancing security, AI has become an indispensable tool for electrical engineers in the telecommunications field. By harnessing the power of AI, telecommunication providers can deliver better services, adapt to changing customer needs, and stay ahead in the ever-evolving digital landscape.

In conclusion, this subchapter on "Artificial Intelligence in Telecommunications" provides an overview of how AI is transforming the telecommunications industry and its relevance to the field of electrical engineering. It showcases the various applications of AI, including virtual assistants, network optimization, security, and more. By understanding and embracing the potential of AI, electrical engineers can contribute to the advancement of telecommunications engineering and shape the future of the industry.

Virtual and Augmented Reality in Telecommunications

Virtual and augmented reality (VR/AR) technologies have revolutionized various industries, and telecommunications is no exception. In this subchapter, we will explore the applications and benefits of VR/AR in the field of telecommunications, providing a comprehensive understanding of how these technologies are transforming the way we communicate.

Virtual reality is an immersive technology that creates a simulated environment, allowing users to interact with computer-generated content. Augmented reality, on the other hand, overlays digital information onto the real world, enhancing the user's perception and experience. Together, these technologies have the potential to enhance telecommunications services in numerous ways.

One of the key applications of VR/AR in telecommunications is in remote communication and collaboration. With VR/AR, individuals can interact with each other in virtual environments, regardless of their physical location. This enables real-time collaboration, making it easier for teams to work together, even when they are geographically dispersed. For example, engineers can use VR/AR to remotely troubleshoot technical issues, reducing the need for costly on-site visits.

Another area where VR/AR is making significant strides in telecommunications is customer service. Through VR/AR, service providers can provide virtual assistance to customers, guiding them through troubleshooting processes or offering product demonstrations. This improves customer satisfaction and reduces the need for physical support visits, resulting in cost savings for service providers.

In addition to communication and customer service, VR/AR is also revolutionizing the way we experience telecommunications services. For instance, with VR headsets, users can immerse themselves in virtual environments and enjoy a more interactive and engaging entertainment experience. This opens up new possibilities for content providers and telecommunication companies to deliver unique and personalized services to their customers.

Moreover, VR/AR is being used in the design and planning of telecommunication networks. By creating virtual models of physical infrastructure, engineers can test and optimize network configurations before implementation, reducing the risk of errors and improving overall efficiency.

In conclusion, virtual and augmented reality technologies have immense potential in the field of telecommunications. From remote collaboration and customer service to enhancing user experiences and network design, VR/AR is transforming the way we communicate and interact in the digital age. As electrical engineering professionals, it is crucial to stay updated with these advancements to leverage their benefits and contribute to the evolution of the telecommunications industry.

Chapter 9: Telecommunications Engineering Careers and Professional Development

Telecommunications Engineering Roles and Responsibilities

In the fast-paced world of telecommunications, the role of a telecommunications engineer is indispensable. This subchapter aims to provide a comprehensive overview of the various roles and responsibilities that telecommunications engineers undertake. Whether you are an aspiring telecommunications engineer, an electrical engineering student, or simply interested in the field, this information will prove valuable to you.

Telecommunications engineers play a vital role in the design, development, and maintenance of communication systems. Their responsibilities encompass a wide range of tasks, including but not limited to:

1. Designing and developing communication networks: Telecommunications engineers are responsible for creating and implementing efficient and reliable communication networks. This involves designing the layout, selecting appropriate equipment, and ensuring seamless connectivity.

2. Conducting feasibility studies: Before implementing any communication network, telecommunications engineers assess the feasibility of the project. They evaluate factors such as cost, technical requirements, and potential challenges to determine if the proposed solution is viable.

3. Network optimization: Telecommunications engineers are constantly monitoring and optimizing network performance. They analyze data traffic, identify bottlenecks, and implement solutions to ensure optimal network utilization.

4. Troubleshooting and maintenance: When issues arise, telecommunications engineers are at the forefront of resolving them. They troubleshoot network problems, identify the root cause, and implement corrective measures to minimize downtime.

5. Upgrading and integrating new technologies: As technology advances, telecommunications engineers are responsible for staying updated with the latest trends. They assess the potential benefits of new technologies and integrate them into existing networks to enhance performance and capabilities.

6. Compliance with regulations and standards: Telecommunications engineers must ensure that communication networks comply with industry regulations and standards. This includes aspects such as data security, privacy, and network reliability.

7. Collaboration with stakeholders: Telecommunications engineers often work closely with various stakeholders, including clients, vendors, and technicians. Effective communication and collaboration are essential to successfully deliver projects and meet client requirements.

By mastering these roles and responsibilities, telecommunications engineers contribute to the advancement of electrical engineering and play a crucial role in enabling seamless communication worldwide. Whether it's designing cutting-edge communication networks or troubleshooting complex issues, the expertise of telecommunications engineers is essential in today's interconnected world.

In conclusion, this subchapter provides a comprehensive overview of the roles and responsibilities of telecommunications engineers. From designing and developing communication networks to troubleshooting and maintenance, telecommunications engineers play a vital role in the field of electrical engineering. By understanding and mastering these responsibilities, aspiring engineers can excel in their careers and contribute to the ever-evolving world of telecommunications.

Educational and Certification Requirements

In the rapidly evolving field of telecommunications engineering, staying ahead of the curve is essential. To succeed in this dynamic industry, individuals must possess the necessary educational background and certifications. This subchapter will outline the educational requirements and certifications that aspiring telecommunications engineers should consider to excel in their careers.

Education is the foundation of any successful career, and telecommunications engineering is no exception. A strong educational background in electrical engineering serves as an excellent starting point for individuals looking to enter this field. A bachelor's degree in electrical engineering or a related discipline provides a solid foundation in the principles and theories of telecommunications engineering. Courses in digital communication systems, signal processing, network architecture, and wireless communications are particularly valuable.

However, the pursuit of higher education does not end with a bachelor's degree. To truly excel in the field of telecommunications engineering, individuals should consider obtaining a master's degree or even a Ph.D. in electrical engineering. These advanced degrees provide a more specialized knowledge base and open up opportunities for research and development in cutting-edge technologies.

Certifications also play a crucial role in the telecommunications engineering industry. One such certification that is highly regarded is the Certified Telecommunications Network Professional (CTNP) certification. This certification validates an individual's expertise in designing, implementing, and managing telecommunications networks. It covers a broad range of topics, including network architecture, protocols, security, and performance optimization.

Another certification that holds significant value is the Cisco Certified Network Professional (CCNP) certification. This certification focuses on the design and implementation of Cisco network solutions, making it particularly relevant for individuals interested in working with Cisco technologies.

Furthermore, industry-specific certifications, such as certifications in wireless communication technologies or satellite communications, can provide a competitive edge in the job market. These certifications demonstrate a candidate's specialized knowledge and expertise in specific areas of telecommunications engineering.

In conclusion, a solid educational background in electrical engineering coupled with relevant certifications is crucial for success in the field of telecommunications engineering. Aspiring engineers should pursue a bachelor's degree in electrical engineering and consider furthering their education with a master's degree or Ph.D. Additionally, certifications such as CTNP and CCNP, as well as industry-specific certifications, can enhance an individual's credentials and open doors to exciting opportunities in the telecommunications industry. By investing in education and certifications, individuals can position themselves for a thriving career in this ever-evolving field.

Professional Organizations and Networking Opportunities

In the ever-evolving field of telecommunications engineering, it is crucial for professionals to stay connected and up-to-date with the latest advancements, industry trends, and networking opportunities. This subchapter explores the importance of professional organizations and networking in the context of electrical engineering and offers insights on how they can benefit individuals in this niche.

Professional organizations play a vital role in shaping and advancing the field of electrical engineering. These organizations bring together like-minded professionals, researchers, and academics to share knowledge, collaborate on projects, and discuss emerging technologies. By joining a professional organization, individuals gain access to a vast network of experts who can provide guidance, mentorship, and career advice.

One such prominent organization in the electrical engineering niche is the Institute of Electrical and Electronics Engineers (IEEE). With numerous specialized societies and technical councils under its umbrella, IEEE offers a wealth of resources, conferences, journals, and networking opportunities. By becoming an IEEE member, electrical engineers can connect with peers, attend conferences, and access cutting-edge research and publications in their field.

Apart from IEEE, there are also other organizations catering to specific areas within electrical engineering. For instance, the Power Engineering Society (PES) focuses on advancements in power systems, while the Communications Society (ComSoc) specializes in telecommunications and networking. By joining niche-specific organizations, professionals can dive deeper into their area of interest and build connections with experts in their field.

Networking opportunities provided by professional organizations are invaluable for career growth. Attending conferences, workshops, and seminars allows individuals to meet industry leaders, learn about the latest developments, and showcase their expertise. These events often feature panel discussions, keynote speeches, and technical sessions that foster knowledge exchange and collaboration.

Furthermore, professional organizations frequently host job fairs and career expos, connecting individuals with potential employers and opening doors to new career opportunities. By actively participating in these events, electrical engineers can enhance their visibility in the industry and stay ahead of the curve.

In conclusion, professional organizations and networking opportunities are essential for electrical engineers looking to excel in their field. By joining these organizations, individuals gain access to a vast network of professionals, valuable resources, and career advancement opportunities. Through networking events, conferences, and workshops, professionals can expand their knowledge, collaborate with experts, and explore new career prospects. In the fast-paced world of telecommunications engineering, staying connected and engaged with professional organizations is a key step towards mastering the field.

Career Advancement Pathways

In the rapidly evolving field of telecommunications engineering, having a clear and defined career advancement pathway is crucial for professionals looking to excel and make significant contributions. This subchapter aims to provide guidance and insights into the various career advancement pathways available in the field of telecommunications engineering, particularly targeted towards individuals with a background in electrical engineering.

1. Technical Specialization: One of the most common career advancement pathways in telecommunications engineering is through technical specialization. By focusing on a specific area of expertise, such as wireless communications, network design, or fiber optics, professionals can become subject matter experts and sought-after specialists in their respective fields. This pathway involves continuous learning and staying abreast of the latest technological advancements to remain at the forefront of the industry.

2. Project Management: For individuals who enjoy leading teams and overseeing complex projects, a career advancement pathway in project management is a viable option. Telecommunications engineering projects often require coordination between various stakeholders, including engineers, technicians, and clients. Professionals with strong organizational, communication, and leadership skills can excel in this pathway, taking on roles such as project managers, program managers, or even department heads.

3. Research and Development:
For those with a passion for innovation and pushing the boundaries of technology, a career advancement pathway in research and development (R&D) can be highly rewarding. This pathway involves conducting cutting-edge research, discovering new solutions, and developing advanced telecommunications technologies. Positions in R&D can range from research engineers to technology managers, working either in academia or industry.

4. Consulting and Entrepreneurship:
Telecommunications engineering professionals with a strong entrepreneurial spirit can explore career advancement pathways in consulting or starting their own ventures. Consulting allows individuals to leverage their expertise to provide advisory services to clients, while entrepreneurship offers the opportunity to develop and commercialize innovative telecommunications solutions. These pathways require a combination of technical knowledge, business acumen, and networking skills.

Regardless of the chosen career advancement pathway, continuous professional development is essential. This can be achieved through pursuing advanced degrees, attending industry conferences, obtaining relevant certifications, and participating in professional associations. Networking and building connections within the telecommunications engineering community also play a crucial role in career advancement.

In conclusion, the field of telecommunications engineering offers a multitude of career advancement pathways for individuals with a background in electrical engineering. Whether one chooses to specialize technically, venture into project management, engage in research and development, or explore consulting and entrepreneurship, the key to success lies in continuous learning, staying updated with technological advancements, and networking with industry professionals. By following these pathways, individuals can forge successful careers and make significant contributions to the ever-evolving world of telecommunications engineering.

Chapter 10: Telecommunications Engineering Case Studies

Case Study 1: Design and Implementation of a Wireless LAN

In the rapidly evolving world of telecommunications, the design and implementation of wireless local area networks (LANs) have become an integral part of electrical engineering. This case study delves into the intricacies of creating a robust and efficient wireless LAN system, providing a comprehensive guide for professionals and enthusiasts alike.

Wireless LANs have revolutionized the way we connect and communicate, offering the freedom and flexibility that traditional wired networks cannot provide. This case study explores the various aspects involved in designing and implementing a wireless LAN, taking into account factors such as coverage, capacity, security, and scalability.

The chapter begins by introducing the fundamental concepts of wireless LANs, including protocols, frequency bands, and modulation techniques. It elaborates on the advantages and limitations of wireless technology, providing insights into the challenges faced during the design process.

Next, the case study focuses on the planning phase, which involves analyzing the requirements and constraints specific to the deployment environment. It discusses techniques for site surveying, signal propagation modeling, and interference analysis. Additionally, it highlights the importance of considering factors such as data rates, network topologies, and access point placement to ensure optimal performance.

The implementation phase is then explored, guiding the reader through the process of configuring access points, setting up security mechanisms, and fine-tuning the network for optimal performance. It explains various security considerations, such as authentication, encryption, and intrusion detection, to safeguard the wireless LAN from potential threats.

The case study also addresses the challenges associated with the ongoing management and maintenance of a wireless LAN. It emphasizes the significance of monitoring network performance, troubleshooting connectivity issues, and upgrading the network to accommodate future technological advancements.

Throughout the chapter, real-world examples and practical insights are provided, enabling readers to gain a deeper understanding of the intricacies involved in designing and implementing wireless LANs. Additionally, emerging trends and technologies, such as the Internet of Things (IoT) and 5G, are discussed, showcasing the evolving landscape of wireless telecommunications.

Overall, this case study serves as a comprehensive guide for electrical engineering professionals, as well as anyone interested in wireless LAN design and implementation. By delving into the various aspects of wireless technology, it equips the reader with the knowledge and tools necessary to create robust and efficient wireless LAN systems in today's interconnected world.

Case Study 2: Fiber Optic Network Deployment for a City

In this case study, we will explore the successful deployment of a fiber optic network for a city, highlighting the key considerations, challenges, and benefits associated with such a project. Fiber optic technology has revolutionized the field of telecommunications, offering high-speed data transmission and exceptional bandwidth capabilities. This case study serves as an excellent example of how electrical engineering principles can be applied to transform the connectivity landscape of an entire city.

The initial step in this deployment was to conduct a comprehensive survey of the city's existing infrastructure, including the assessment of existing copper or coaxial cables. This survey helped in identifying areas that required fiber optic connectivity upgrades and determining the optimal routes for laying the fiber optic cables. The electrical engineering team worked closely with city officials and stakeholders to ensure that the deployment plan aligned with the city's development goals and met the needs of its residents.

As with any large-scale project, the deployment of a fiber optic network presented several challenges. One significant challenge was securing the necessary permits and rights-of-way for trenching and installing the cables. The team had to navigate through a complex web of regulations and coordinate with various utility companies and local authorities. Additionally, the project required meticulous planning to minimize service disruption to businesses and residents during the installation process.

Despite these challenges, the benefits of deploying a fiber optic network were significant. By replacing outdated copper infrastructure with fiber optic technology, the city experienced a dramatic increase in internet speeds and bandwidth capacity. This upgrade facilitated the expansion of digital infrastructure, enabling the implementation of smart city initiatives, such as improved traffic management systems, smart grid integration, and enhanced public safety networks.

Moreover, the fiber optic network deployment attracted businesses and investors to the city, driving economic growth and creating job opportunities. The enhanced connectivity also improved access to online education, healthcare services, and e-commerce platforms, benefiting residents of all ages and backgrounds.

This case study exemplifies the transformative power of fiber optic technology in revolutionizing a city's telecommunications infrastructure. It underscores the crucial role of electrical engineers in planning, designing, and implementing such projects. By mastering the principles of telecommunications engineering, professionals can contribute to the development of future-proof and resilient connectivity solutions that empower communities and drive economic prosperity.

Case Study 3: Upgrading a Telecommunication Network for a Telecom Provider

Introduction:

In this case study, we will explore the process of upgrading a telecommunication network for a telecom provider. With the rapid advancements in technology and increasing customer demands, it is crucial for telecommunication companies to continuously upgrade their networks to provide better services. This case study focuses on a real-life scenario where a telecom provider successfully upgraded their network, resulting in improved performance and customer satisfaction.

Background:

The telecom provider in question had been facing numerous challenges due to an outdated network infrastructure. The network was unable to handle the increasing volume of data traffic, resulting in frequent dropped calls, slow internet speeds, and poor service quality. To address these issues, the provider decided to upgrade its telecommunication network.

Upgrading   the   Network:

The first step of the upgrade process involved conducting a comprehensive analysis of the existing network infrastructure. This included assessing the network capacity, identifying bottlenecks, and determining the areas that required immediate attention. The telecom provider collaborated with experienced electrical engineers who specialized in telecommunication networks to perform this analysis.

Based on the analysis, the telecom provider developed a detailed plan for upgrading the network. The plan included deploying advanced routers and switches, upgrading the backbone connectivity, implementing fiber-optic cables, and enhancing the network security measures. The electrical engineers worked closely with the telecom provider's team to ensure the smooth execution of the upgrade plan.

Benefits   and   Results:
After the successful network upgrade, the telecom provider experienced several significant benefits. The upgraded network offered faster internet speeds, improved call quality, and enhanced reliability. The new infrastructure also allowed the telecom provider to handle the increasing data traffic more efficiently, resulting in a reduction in dropped calls and network congestion.

Additionally, the upgraded network enabled the telecom provider to introduce new services and features to its customers. These included high-definition video calls, faster download and upload speeds, and seamless connectivity across various devices. As a result, the telecom provider witnessed a significant increase in customer satisfaction and loyalty, leading to a surge in their subscriber base.

Conclusion:
The case study highlights the importance of upgrading telecommunication networks for telecom providers. By investing in network infrastructure and collaborating with experienced electrical engineers, telecom providers can overcome challenges and deliver improved services to their customers. The successful network upgrade resulted in enhanced performance, increased customer satisfaction, and a competitive advantage for the telecom provider.

Chapter 11: Conclusion

Recap of Key Concepts

In this subchapter, we will revisit the essential concepts covered in the previous chapters of "Mastering Telecommunications Engineering: A Comprehensive Guide." Whether you are a student, aspiring engineer, or a seasoned professional in the field of electrical engineering, this recap will serve as a useful refresher and reference for the key principles and techniques discussed throughout the book.

Firstly, we introduced the fundamental concepts of telecommunications, including the transmission of information through various mediums such as wires, cables, and wireless technologies. We explored the basics of signal processing, modulation, and demodulation techniques, which are crucial for transmitting and receiving data effectively.

Next, we delved into the world of data communication protocols and networking. We discussed the OSI (Open Systems Interconnection) model, TCP/IP (Transmission Control Protocol/Internet Protocol), and various network topologies such as LAN (Local Area Network) and WAN (Wide Area Network). Understanding these concepts is essential for designing and implementing efficient communication systems.

The subchapter also covered the principles of digital communication, including encoding, decoding, and error detection and correction techniques. We explored various digital modulation schemes, such as amplitude shift keying (ASK), frequency shift keying (FSK), and phase shift keying (PSK), as well as the concept of multiplexing to maximize the utilization of communication channels.

Furthermore, we discussed the basics of telephony systems, including analog and digital voice transmission, voice codecs, and the integration of traditional telephony with IP-based networks. We also touched upon the latest trends in telecommunications, such as Voice over IP (VoIP), virtualization, and cloud-based services.

Throughout the book, we emphasized the importance of understanding the underlying principles of telecommunications engineering. By grasping the key concepts, you will be able to design, implement, and troubleshoot complex communication systems confidently.

Whether you are interested in working on advanced wireless technologies, network infrastructure, or the development of cutting-edge communication devices, this book has provided you with a solid foundation. By mastering the key concepts outlined in this subchapter, you will have the necessary knowledge and skills to excel in the field of electrical engineering and make meaningful contributions to the telecommunications industry.

In conclusion, the "Recap of Key Concepts" subchapter serves as a valuable summary of the essential topics covered in "Mastering Telecommunications Engineering: A Comprehensive Guide." It provides a comprehensive overview of the principles and techniques discussed throughout the book, enabling professionals and aspiring engineers in the field of electrical engineering to reinforce their understanding and apply these concepts effectively in their respective fields.

Future Directions in Telecommunications Engineering

Telecommunications engineering has come a long way since the invention of the telephone by Alexander Graham Bell in 1876. Over the years, the field has witnessed remarkable advancements, transforming the way we communicate and connect with each other. As technology continues to evolve at an unprecedented pace, it is crucial to explore the future directions in telecommunications engineering to understand the potential impact on our lives and the opportunities it presents.

One of the most exciting areas in telecommunications engineering is the development of 5G technology. 5G promises to revolutionize the way we use the internet, enabling faster speeds, lower latency, and greater capacity. This will pave the way for innovative applications such as autonomous vehicles, remote surgery, and augmented reality. Telecommunications engineers will play a pivotal role in designing and implementing the infrastructure required to support the widespread adoption of 5G.

In addition to 5G, the Internet of Things (IoT) is another area that holds great promise for telecommunications engineering. IoT refers to the network of interconnected devices and objects that can communicate and share data with each other. From smart homes to industrial automation, the potential applications of IoT are vast. Telecommunications engineers will be at the forefront of developing secure and reliable communication protocols to facilitate seamless connectivity among these devices.

Artificial Intelligence (AI) is yet another area that will shape the future of telecommunications engineering. With AI, telecommunication networks can become smarter and more efficient, optimizing resource allocation and predicting network failures. Telecommunications engineers will need to acquire skills in machine learning and data analytics to leverage the power of AI in improving network performance and delivering a better user experience.

As the world becomes increasingly interconnected, cybersecurity will be a pressing concern for telecommunications engineers. The rise in cyber threats and attacks poses significant challenges in ensuring the integrity and confidentiality of communication networks. Telecommunications engineers will need to develop robust security measures and protocols to protect against unauthorized access and data breaches.

In conclusion, the future of telecommunications engineering holds immense potential for innovation and growth. From the advent of 5G and IoT to the integration of AI and the increasing importance of cybersecurity, this field will continue to shape the way we communicate and interact with technology. Aspiring telecommunications engineers should stay informed about these future directions and equip themselves with the necessary skills and knowledge to contribute to this dynamic and rapidly evolving field.